What's Your Dog's IQ?

How to Determine if Your Dog Is an Einstein—and What to Do if He's a Scooby Doo

Sue Owens Wright

Adams Media
Avon, Massachusetts

Published by Adams Media, an F+W Publications Company
57 Littlefield Street
Avon, MA 02322
www.adamsmedia.com

ISBN: 1-59337-602-2

Printed in Canada.
J I H G F E D C B A

Library of Congress Cataloging-in-Publication Data
Wright, Sue Owens.
What's your dog's IQ?/Sue Owens Wright.
p. cm.
ISBN 1-59337-602-2
1. Dogs--Psychology--Testing. 2. Animal intelligence--Psychology--Testing. I. Title.
SF433.W77 2006
636.7--dc22
2006005003

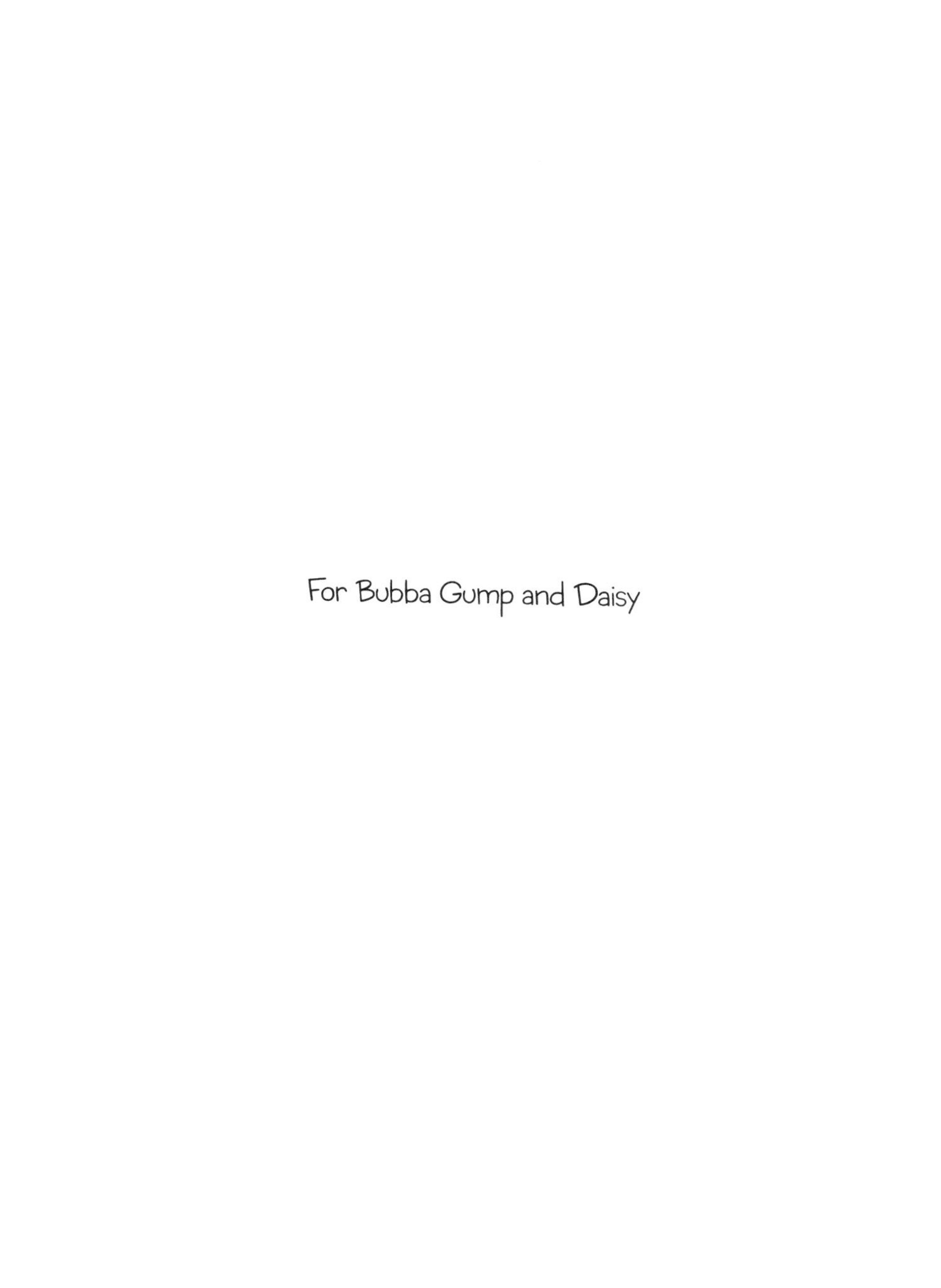

For Bubba Gump and Daisy

Contents

Acknowledgments

Thanks to my wonderful agent, editors, and publisher, without whom I would not have had the opportunity to write this book. Thanks also to the many contributors who generously shared their own tales of intelligent dogs. Last, but not least, thanks to my dear mother, who always made sure that I had a dog to love.

Introduction

"I've seen a look in dogs' eyes, a quickly vanishing look of amazed contempt, and I am convinced that basically dogs think humans are nuts."—John Steinbeck

Sure, all dogs have doggie smarts, but just how *smart* is your dog? Is he as smart as Eddie, the Jack Russell terrier made famous on the television sit-com *Frasier*? If so, you could be living with your own pup prodigy. But maybe your dog's more like Satchel, the dazed and confused shar-pei from the cartoon strip *Get Fuzzy*, or Quincy, the laid-back basset hound from the television series *Coach*. If your dog is like most, he doesn't fall so clearly or easily into either category. This book explores various aspects of the canine intellect. Some of these are measured by tests, while others can be measured only by the heart.

Starting Young: When We Fell In Love with the "Smart Dog"

As children, we loved smart dogs. We all yearned for dogs that could perform their pet duties while also throwing in the extra heroism that made them come close to the genius dogs we saw in the movies and on television. We adored Lassie, the beautiful, intelligent collie who always alerted Timmy's folks when he had fallen in the well. She barked and beckoned until his parents got a clue that they should follow her and bail him out again. She was almost smart enough to talk. If only she had been able to! Maybe she would have convinced Paul and Ruth to cover the well so that stupid Timmy wouldn't keep falling in.

There was also Rin-Tin-Tin, or Rinty, as he was affectionately known to his pint-sized fans, the smart matinee idol German Shepherd that bit bandits and fended off Apaches. And many of you probably remember Bullet, Roy Rogers' savvy, speedy sidekick, who could run even faster than Trigger the Wonder Horse.

Old Yeller was also a dog to be admired for his intelligence and canine cunning. Here was a thinking dog that could really puzzle things out. When he cocked his head at Little Arliss in that certain way, you knew he was deeply immersed in solving a difficult problem, like how to get to the smoked ham hocks that were hung high enough on the cabin porch to keep them out of his reach (or so his people thought).

Then there's good ol' Snoopy, Charles Schultz's cartoon canine, who could not only pilot an airplane and dance like Fred Astaire but was also writing a novel, even if it did have the worst opening line in the history of literature: "It was a dark and stormy night."

An enduring Saturday morning kid's cartoon favorite is Scooby Doo. Even though he's admittedly a scaredy dog, Scooby can walk and talk and has a knack for ghost busting and crime solving. Even better, he's not the only Sherlock Bones in the world of fantasy and fiction. Some dogs have even served as secret agents, like Fang, the Labradoodle that played Agent K-13 in the classic comedy series *Get Smart.* That most lovable mutt, Benjy, has also been known to do a little sleuthing now and then.

Most children could tell you all about the adventures of Wishbone, a bright and lively Jack Russell terrier with the performing skills of a Shakespearean actor. He's even portrayed the Bard himself. If you're in search of dogs of literary distinction, look no further than John Dufresne's most recent novel, *Johnny Too Bad,* which features a dog named Spot—the name of another classic smart dog many of us remember from when we first learned how to read. "See Spot run. Run, Spot, run!"

My dog Dusty was the kind of dog that would jump into a river to save someone from drowning. Once he did exactly that. One hot summer day, my mother was swimming in the river, splashing around and squealing. Thinking she was drowning, Dusty leaped in and swam straight for her. Brow furrowed with worry, he dog-paddled for all he was worth, intent on the task at hand. If Mom really had been in distress, Dusty would have been a hero that day. He was always a hero to me.

It wasn't just kids who jumped on the bandwagon. Intelligent dogs were so highly regarded that in the 1960s, a dog-food manufacturer produced a catchy commercial jingle about the ongoing competition over having the smartest dog: "My dog's smarter than your dog; my dog's smarter than yours. My dog's smarter 'cause he eats Kennel-Ration; my dog's smarter than yours." Maybe he was smarter, and maybe he wasn't. No one could prove that feeding your dog Kennel-Ration or any other brand of dog food would make him smarter, but I persuaded my parents to buy only Kennel-Ration for our dogs, just in case the jingle was true.

Intelligent dogs are more trainable, and people naturally prefer having a dog that's easy to train and that will respond to their commands. A bright, well-behaved dog is just more entertaining and pleasant to be around. You and your family are more likely to enjoy the company of an obedient and engaging dog, and you can feel assured that others will enjoy a smart dog's company, too.

It's understandable for a new pet owner to want the best dog of the litter. Choosing a dog is the only opportunity you'll ever have to choose a relative, and most people would rather choose one who is smart, not dumb. Just as parents do with their children, people have a tendency to think of their dogs as extensions of themselves. You wouldn't want to think your child was a flunky or see him jumping all over the furniture or biting someone else's child. Our dogs are part of our family, so naturally they are a reflection on us.

Our dogs are also a source of pride. We talk endlessly about our dogs with other doting dog owners and love to brag about them to anyone who'll listen without gagging. Some owners even have photos handy in their wallets and purses, digital cameras, or cell phones, ready to whip out and share with anyone who comes within leash length. Proud pup parents expect their candid canine camera shots to elicit as many "Oohs" and "Ahhs" among dog lovers as they would get if they were sharing photos of human children, which is usually the case unless the person doing the looking is actually a cat fancier.

There's obviously no genetic family resemblance between us and our dogs—although you may have noticed some strange look-alikes in some cases. Because they are an extension of our family, we want to believe that our canines are the cleverest since Pavlov's dog first drooled on command.

In the past, there were no tests available for dog owners to determine whose dog was the smartest. No one, not even scientists, knew or cared much about how to measure intelligence in dogs. (Even Pavlov wasn't really measuring intelligence; all he really did was time digestive responses in his dogs.) They've been busy with more important things, like building atomic bombs and landing on the moon. Scientists are still building atomic bombs, but now they have also finally devised a way to find out if your dog is smarter than the average dog. Keep reading . . . if you're sure you really want to know.

As you read these chapters and begin to test your dog's intelligence, it may be hard not to stack his performance up against some real or fictional Superdog of your past. You might also be tempted to compare him to your neighbor's dog, the one that fetches the daily paper, collects the mail, and turns off the light switch whenever he leaves a room. Even if your dog never won Best in Show at Westminster, performed in the Ringling Brothers Circus, or earned a Ph.D. in Dogology, he certainly excels at other amazing feats. He wants to be in your company constantly, agrees with everything you say, and greets you like a returning war hero, even if you've only been out of the room for five minutes. Can you say that about anyone else, even in your own family?

So lighten up, and have some fun with the exercises and tests in the book. Make them fun for your dog, too. Remember that he's not competing for a blue ribbon here, just a doggie treat or a special pat from you. Not only will you enjoy some quality time with your best friend, he might surprise you. You could discover that you have a smarter dog than you thought. You might even have a budding Lassie on your hands. Keep in mind that you might also discover he's just a clueless Scooby Doo. If so, don't hold it against him or withhold the treats. He deserves to be showered with Snausages. After all, he's still the best there is at the most amazing feat of all, one that humans haven't mastered and probably never will—giving unconditional love.

That's one test for your dog you won't find in this book, for the very good reason that it hasn't been invented yet. There's just no way to accurately measure the most important part of the dog: the heart.

Chapter 1

Why Test Your Dog's Intelligence?

"A dog teaches a boy fidelity, perseverance, and to turn around three times before lying down."—Robert Benchley

The quest for the smartest dog does not stop at childhood. Even as grownups, the competition endures between us over who has the smartest dog, only now it's "My fur kid's smarter than your fur kid." A large number of people live alone. For many of those, dogs have become the sole beneficiaries of their love and attention. In other cases, where a house may be bustling with one or several children, a dog can serve as a playmate that ends up requiring almost as much time as one of the kids. Just ask any mom who finds herself walking the "family" dog on those snowy mornings, while the children who promised to walk and feed the dog every day are still snuggled in their beds sleeping soundly. Chances are that Mom even takes time to outfit the dog in his designer parka and snow booties before she walks him. Regardless of whether a dog is a sole companion or a third ring to the family circus, one thing is for sure: Americans are spending much of their time—and money—on their dogs.

> "If you are a dog and your owner suggests that you wear a sweater . . . suggest that he wear a tail."—Fran Lebowitz

America's Prissy, Pricey Pups

According to statistics compiled by the American Pet Products Manufacturers Association (APPMA), 65 million dogs are kept as pets in American households. Pet spending has more than doubled over the past decade, from $17 billion in 1994 to $35.9 billion in 2005. According to the U.S. Census Bureau, the pet industry is now the seventh-largest retail segment in the country. That $35.9 billion spent in 2005 breaks down approximately as follows:

- $14.5 billion for food
- $8.8 billion for supplies and over-the-counter medications
- $8.6 billion for veterinary care (excluding prescriptions)
- $1.6 billion for live animal purchases
- $2.4 billion for other services

That's a whole lot of pampering and pup paraphernalia. Clearly, nothing is too good for our dogs. Logically enough, we want these fur-bearing creatures to be worthy of all the time, money, and attention we lavish upon them, and that means we want them to be considered intelligent by the experts. If President Bush had a "Leave No Dog Behind" program, we would probably be pressuring our fur kids to excel and pass tests before they could graduate from puppy kindergarten.

"My dog is worried about the economy because Alpo is up to 99 cents a can. That's almost $7 in dog money."—Joe Weinstein

Dog parents are becoming as competitive as parents of children. If you don't believe me, attend any dog show or agility or obedience competition and observe the owners. Soccer moms and Little League dads have nothing on this overzealous bunch. Yet if you take another look at that list of doggie spending categories, you might notice that doggie intelligence was conspicuous only by its absence. We want out dogs to be smart, but we don't spend the money.

"When a man's best friend is his dog, that dog has a problem."—Edward Abbey

For many of us, our dogs are our surrogate children. We take them everywhere we go. We dress them up like miniatures of ourselves, which they must hate. We spend a small fortune on products to make their lives more comfortable and enjoyable. We send them to doggie day spas or to the acupuncturist for their health. Some people are even willing to get cosmetic surgery for their dogs, a rather extreme measure—not that I haven't seen a basset hound or two that could use a facelift.

We call in the dog whisperer, animal behaviorist, or pet psychic when Fido has behavioral or emotional issues. (You have to wonder what past-life regression for a dog must be like.) We feed them special foods, clean up after them, and jump up whenever they bark to be let in or out, regardless of whether we've provided them with their own doggie doors. One might ask just who has whom well trained and which creature is the more intelligent: the one with two legs or four legs?

Perhaps this species that has allied itself so closely to humans is far more intelligent than we give credit for.

> "Do not make the mistake of treating your dogs like humans, or they will treat you like dogs."—Martha Scott

Different Kinds of Dog Smarts

Many different elements enter into determining a dog's intelligence. Is the dog observant? How good is the dog at adapting to its environment? How well does he understand verbal or hand commands? Does he learn new tasks easily? How's his long- and short-term memory? The problem becomes apparent almost right away, and that has to do with objectivity. How do you judge the answers to these questions with any degree of exactitude? No matter how many elements you test or how many characteristics you come up with, determining canine intelligence is not an exact science.

Still, we can take a stab at it. According to Stanley Coren, author of *The Intelligence of Dogs,* there are three types of dog intelligence:

- Adaptive intelligence (learning and problem-solving ability)
- Instinctive intelligence (individual intelligence)
- Working/obedience intelligence (breed-dependent intelligence)

Let's take a closer look at these types of smarts, so you can try to figure out where your pup fares best.

Adaptive Intelligence

Adaptive intelligence refers to your dog's ability to learn new tasks and solve problems. How well does he understand the laws of cause and effect? For instance, he probably knows that if he scratches at the back door long enough, you'll eventually get off your duff and let him out. Or if he stares at you and whines incessantly, you'll give in and share a bite of your T-bone with him. If he's also figured out how to use the remote control to tune into Animal Planet, you can bet he's off the charts in terms of adaptive intelligence—either that or Nielsen has been polling your pup when you're not around.

Beyond Book Smart

The largest litter of puppies ever birthed by one pup—twenty-four Neopolitan mastiffs—was born to mother Tia on November 29, 2004. So don't feel badly if your dog doesn't ace his SATs—Tia admits she rarely has time to help her kids study!

How long does it take your dog to adapt to new situations or new relationships? Does your dog recognize guests in your home after just one or two visits? If he does, he'll score high on the adaptive intelligence scale. However, if he still bares his teeth at your boyfriend or pees on his pants leg even though you've been dating the guy for five years, then your dog probably doesn't score too high on the adaptive intelligence scale. Then again, maybe he's an excellent judge of character and is just trying to tell you something about your choice in men.

"Some days you're the dog; some days you're the hydrant."
—Unknown

Probably the greatest proof of the dog's amazing adaptive ability is seen in stray dogs that are adopted from pounds and shelters. These dogs have already shown their capabilities of adaptive intelligence by having survived the mean streets long enough to be apprehended by the dog-catcher. On top of that, they have to adjust to life with strangers in a different house, maybe even in a different town or another country. If that's not adaptive intelligence, I don't know what is (even though it may take them a while to break in their new owners).

Instinctive Intelligence

Instinctive intelligence includes the behaviors and skills imprinted in your dog's genetic code. Dogs have a propensity to do doggie things, like lift a leg on the shrub to mark territory, bury a bone to hide it from competitors, or circle three times before lying down. (The jury's still out on the reason for this strange behavior, but it's probably to flatten down their bedding or to check for danger before sleeping, though it could also be some sort of doggie obsessive-compulsive thing, like excessive hand-washing in humans.) Whatever the case, dogs will be dogs. Their canine programming gives them no choice but to exhibit the particular universal doggie behaviors that are imprinted in their genes and that help them survive and produce litters of more dogs.

Beyond Book Smart

The smallest living dog, in terms of length, is a long-haired chihuahua named Danka Kordak Slovakia, who measures in at only 5.4 inches. Does he even have room for brains?!?!

To take the idea of genes even further, you'll also find that a dog's behavior differs from that of other dogs depending on its breeding. All dogs possess certain traits that are instinctive to their breed (or combination thereof). Sheepdogs and Border collies, for instance, are sheep-herding fools. Labrador retrievers will retrieve just about anything retrievable, and bloodhounds will follow a scent until it or the dog gives out. That's what they were bred to do. With proper training, these skills can all be honed even more sharply, but if you're not careful you can end up with an obsessive-compulsive canine.

So if you don't have a sheep ranch for your collie, a lake for your Lab, or plenty of sniffable acres for your scent hound, proceed with caution.

The worst mistake a potential owner can make, and the reason so many dogs end up in shelters and pounds, is to fail to consider the instinctive intelligence factor when choosing a dog. The dog should fit your lifestyle. It's just canine common sense.

If you're a long-distance runner, you probably won't be happy with a slo-mo basset hound and its short stubby legs; if you're a couch spud, forget about owning an energetic herding breed like the Australian cattle dog. You'll both be miserable.

Working/Obedience Intelligence

Working/obedience intelligence is all about how well a dog obeys commands. The breed of the dog comes into play here, too, as well as its environmental, social learning, and language comprehension capabilities. It all boils down to the fact that some breeds of dogs are naturally more obedient and trainable than others and better suited to performing certain tasks. That's why you'll probably never see a scent hound used as a Seeing Eye dog or a Pekingese cruising in the back seat of a police K-9 unit. At the same time, training and the desire to please are also very important. If you've never studied astronomy or rocket science or gone to Space Camp, you probably wouldn't be a good candidate to apply for a career as an astronaut. Of course, Leika the canine cosmonaut never studied rocket science, either, and she was the first dog to travel in space.

Beyond Book Smart

Talk about a show-off . . . Chanda-Leah, a toy poodle, holds the world record for having the largest repertoire of tricks. Her magic number? 469. Who **knows** what her I.Q. is!

Personality differences such as excitability or calmness make dogs better at doing some jobs than others. For instance, some breeds make excellent watchdogs because of their size, aggressiveness, and territory guarding tendencies. Rottweilers, German shepherds, mastiffs, and Doberman pinschers are a few of the breeds commonly used for guarding property, although I once encountered an attack Chihuahua. It also helps if the dog is apt to vocalize to alert its owner to intruders, so that's why you'll probably never see a Basenji, the African barkless dog, guarding a junkyard. (They may not bark, but they can let out a screech that will shatter a windshield. Now that's some subwoofer!)

Sign on fence: "Salesmen Welcome . . . Dog food is expensive"

Despite all our concern about quantifying a dog's adaptive, instinctive, and working intelligence, the only thing we can accurately measure about this species is how much they enrich our lives with their immeasurable capacity for love and devotion. Then again, there are some things that dogs will *never* be able to do.

Top Ten Reasons Dogs Don't Use Computers

10. T0o0p hqa5rxd 6tt0[o 6ty[p3e 2w9igtjh;pa3wds (Too hard to type with paws)

9. "Sit" and "Stay" were hard enough; "Delete" and "Save" are out of the question.

8. Saliva-coated floppy disks refuse to work.

7. Carpal paw syndrome

6. Involuntary tail wagging is dead giveaway he's browsing *www.purina.com*

5. Fire-hydrant icon is simply frustrating.

4. Can't help attacking the screen when he hears, "You've got mail."

3. Too messy to mark every Web site he visits.

2. Fetch command not available on all platforms.

1. Can't stick his head out of Windows XP.

Other Kinds of Doggie Intelligence

While the various aspects of doggie consciousness are certainly part of your dog's intelligence, many other factors also contribute to Fido's intellectual makeup. Dogs are really quite amazing creatures that are similar to us in many aspects of their mental capabilities. Perhaps that's why we get along so well with our canine companions. If you have a dog now, or have ever spent a significant amount of time with one, you have probably noticed some of the following wonderful things about canines:

- Dogs display an array of emotions: They laugh, they cry, they get sad, they get frustrated and angry.
- Dogs get lonely.
- As proved by many of the stories in this book: Dogs can outwit humans!
- We've also seen that dogs can figure stuff out!
- Dogs communicate with us and with each other. They have complex greeting behaviors and watch our body language to gauge our reactions to them.
- Dogs can tell time. They always know when it's time to eat or when we come home!
- Dogs are social creatures. If they are left alone without socialization they become depressed and destructive.
- Dogs have ESP (at least extra sensory compared to us!)

These are just the tip of the T-bone in fully understanding the complexity of a dog's intelligence. If you've lived in the company of dogs as long as I have, you've probably become aware of some of their hidden talents. For instance, do dogs have a sixth sense? I'd swear that my dogs do. Can they see things we can't perceive, just as they can hear things beyond the range of our human hearing? Could you have a metaphysical mutt living in your house? If next Howloween, Madame Pawline begs to use your Ouija board and starts reading the fall leaves in her water bowl, it's a possibility.

One day, while sitting in the front garden with my dogs, I started thinking of Dolly, one of my bassets who had passed on to Rainbow Bridge. I was feeling sad about her untimely loss. Suddenly, there was a gust of wind. My dogs jumped up, reared up on the picket fence, and began barking like crazy at something. I recognized it as their special high-pitched greeting bark that they use for someone they know well. But it wasn't time for their dad to come home, and there was no one else around. I looked up and down the street for any sign of someone. There wasn't a soul there, at least not one that I could perceive.

Operant Conditioning

David Letterman has a segment on his talk show called Stupid Pet Tricks. In my estimation, however, a dog that can count or say, "I love you" to his owner is anything but stupid.

Many owners find that their dogs are more than capable of displaying talents normally reserved for humans. My basset hound Bubba can read a clock; he knows to the minute when it's walkies time and barges into my office promptly every day at 11:00 A.M. and 4:00 P.M. He also has a preference for listening to classical music, even though he's not a college professor. However, he is a wonderful baritone bugler. Next thing I know, he'll probably want his own iDog. Maybe dogs have hung around with humans for so long that they are becoming too much like us through conditioning.

Beyond Book Smart

When they award Best in Show at the Crufts dog show held in Birmingham, England, it really means something. Holding the world record for largest annual dog show, in 1991, it boasted 22,993 canines!

The way a dog learns is not that much different from the way human beings assimilate new information. Like us, a dog learns from the consequences of its behavior and from experience. For instance, your dog discovers that if he drops a ball at your feet, you'll throw it for him, or if he whines to be let out, you'll open the door for him. He learns that some of his actions produce rewards and others bring punishment. That way he knows whether or not to repeat those actions. This is called operant conditioning. According to Stacy Braslau-Schneck, author of *An Animal Trainer's Guide to Operant and Classical Conditioning,* there are four possible consequences to any behavior.

1. Something good can start or be presented.
2. Something good can end or be taken away.
3. Something bad can start or be presented.
4. Something bad can end or be taken away.

With dogs, the negative or positive consequences of their behavior must be immediate in order for them to make the connection the way we humans can. On the other hand, a dog doesn't have to repeat his mistakes nearly as many times to understand he shouldn't do something that results in negative consequences. This is unlike the way most people learn, such as those who refuse to recognize that smoking is detrimental to your health or it's hazardous to drive a car while eating a Krispy Kreme doughnut and talking on a cell phone. Come to think of it, I've never seen a dog do either of those things. Yet.

"A door is what a dog is perpetually on the wrong side of."
—Ogden Nash

Classical Conditioning

Closely related to this kind of operant conditioning is classical, or Pavlovian, conditioning. Named after Russian physiologist Ivan Pavlov, this style of learning involves pairing a stimulus (ringing a bell) with an event (getting fed).

Actually, Pavlov's career began with the study of saliva and the part it plays in the digestive process. It's not everyone who wins the Nobel Prize for studying spit, but in 1904 Ivan Pavlov did.

After winning this claim to fame, Pavlov became interested in the study of reflexes and learned behaviors, for which he is more widely recognized. This study began when Pavlov happened to notice that his dogs drooled predictably even when there was no food present. In other words, the dogs performed the conditioned behavior (drooling) in the absence of the usual stimulus (food). After a little investigation, Pavlov discovered that the dogs were reacting to other stimuli that always signaled the imminence of dinnertime, such as the footsteps of lab technicians or the sight of a lab coat.

He took his research a step further and conducted a series of experiments in which he established a completely unconnected stimulus to the feeding process—the ringing of a bell. The sound of the bell became a pleasant association for the dog and came to mean that dinner would shortly be served. After a while, the dogs drooled at the sound of the bell, regardless of whether they were actually fed.

By the same token, a bad association can be similarly formed, which is why dogs may misbehave at the veterinarian's office or even beforehand, if they sense that is where they are being taken. They probably haven't had too many pleasant experiences there in the past. But at least there's a bell on the office door and the vet wears a lab coat.

Sign at vet's office: "Be Back in 5 Minutes. SIT! STAY!"

Emotional Intelligence

Just as with humans, a dog's emotions play an important part in how he learns. If the dog has had unpleasant or stressful experiences in the past, especially if they were connected to the training process, he may be less able or willing to learn. Harsh correction techniques, such as jerking a dog's collar or hitting him for failing to perform the expected behavior, can have the consequence of killing a dog's desire to perform. He might figure that he tried to do what the trainer wanted and only got punished as a result, so why keep trying? Re-establishing trust is Lesson One, and positive reinforcement works wonders in undoing the bad experiences of the past.

By the same token, you can't teach a dog anything if he's frightened or nervous. Before you can teach your dog anything, you'll need to address his emotional issues, whatever those might be. Anyone who has ever adopted a formerly abused or neglected dog knows exactly what those issues are.

Street Smarts

Some dogs just seem to be able to figure things out on their own—including how to get around in a people's world without a person to help them along. You've probably seen one of these dogs trotting along the street, looking for all the world like he was on the way to the grocery store or an important meeting. They know the shortcuts. They know when it's safe to walk out in the street and when it's a better idea to use the sidewalk. Some people might argue that this is an example of their strong survival instinct at work on the streets. Instead of evading predators in the wild, these dogs have learned it's best to avoid those roaring metal beasts. And what about Seeing Eye dogs, which are trained to stop at intersections and wait for the traffic light to turn green? It makes you wonder whether dogs really can see in color after all (a topic of hot doggie debate).

If you *have* seen a dog waiting for the pedestrian signal to change before crossing the street, you can bet that the dog probably wasn't a basset hound. As I've discovered from having owned six of them over the years, this breed seems to have no road sense whatsoever. Along with any other breed of scent hound, a basset should never be let off-leash near traffic, where it will most likely follow its keen nose right under the wheels of a moving car. It's just the nature of the beast. Scent hounds were bred to track game over long distances. They see the world in smell-a-vision.

"To a dog, the whole world is a smell."—Anonymous

Sight hounds pursue their game by sight and were bred to run it down over great distances. The greyhound is probably the best-known example of such speed, which is why there's a bus company named after them. These dogs will also follow that sighting instinct and dash right onto a busy thoroughfare, especially if he sees a rabbit across the street. No, not a Volkswagen, but he's fast enough to catch that Rabbit, too.

Jaclyn LaDue's rescued greyhound, Katy, was trained to chase a rabbit lure around a racetrack. Jaclyn's neighbor has a Chihuahua that is just about the size of a rabbit. When the Chihuahua came running up to Katy one day, the greyhound picked the little dog up in her mouth. To both owners' relief, Katy must have suddenly decided the Chihuahua wasn't prey, because she promptly spit her "rabbit" out. Katy didn't hurt the little dog, and although it wasn't too funny at the time, when Jaclyn got home, she couldn't help but laugh.

Some dogs also seem to be born map-readers; some can navigate busy roads as though they had built-in GPS technology. The fact so many retrievers are trained as guide or helper dogs attests to their street smarts and their ability to be trained to avoid hazards. Remember *The Incredible Journey*, the story of the two dogs and cat that traveled safely across miles of Canadian wilderness to find their family? If you do, you'll also remember that it was a golden retriever in the lead.

Both golden and Labrador retrievers (two different breeds) are also frequently seen guiding the visually or physically handicapped. These dogs are not only intelligent and highly trainable, but their heritage of spotting birds in the field means they have excellent visual acuity and are keenly aware of sudden movements. Those are good talents to have on busy streets, whether you're dodging a streetcar named Desire or a bus named Greyhound.

Did You Know?

A champion greyhound can run at speeds of approximately forty miles per hour!

Memory

An elephant never forgets, and neither do dogs. My dog, Bubba Gump, remembers every place in the neighborhood where he's ever seen a cat. It's gotten so bad that now, instead of "walks," I call our outings "CAT scans." He also recognizes every house where he's ever been offered a treat.

Other dog owners have been startled by their dog's ability to recall things that occurred in the past. Donna

Deacon's dog Jake used to watch intently as her other dog, Jewel, struggled to extract treats from a Kong (the popular roly-poly dog toy you stuff with tempting goodies, not the giant ape on top of the Empire State Building). Jewel would paw the toy repeatedly with no success before finally picking it up in her teeth and clamping down as hard as she could. Then she'd toss her head back and shake the broken pieces of biscuit into her mouth. A year later, after Jewel was gone, Jake copied her clever trick to get the treats.

We may like to think differently, but dogs never really forget their past experiences or the people they have lived with, particularly if those experiences were in any way traumatic. Just ask anyone who is involved in rescue groups, like Guardian Angel Basset Rescue in the small town of Dwight, Illinois, which has rescued and rehomed over 2,000 basset hounds. Every September they honor their rescuers and rescuees in a celebration called the Illinois Basset Waddle, which draws bassets and their owners from all over the United States, Canada, and even some from abroad. The dogs these kind people take into their homes and hearts frequently come with a whole doggie bag of issues to overcome. But with lots of love, care, and training, the dogs can and do overcome their troubled pasts and go on to live happy, healthy lives with their adoring new families.

Other stories of situations that test a dog's memory have probably been brought to your attention as well. There have been instances where dogs have traveled many miles in search of their former owners. Certainly, instinct and a keen sense of smell are factors in such incredible journeys that we hear about from time to time, but the dog's devotion and the memory of a beloved master cannot be discounted.

"The dog has an enviable mind; it remembers the nice things in life and quickly blots out the nasty."—Barbara Woodhouse

Social Learning

When a pup is born, he doesn't really know that he's a dog. It takes time for him to learn what he needs to know about dogdom from his mother and littermates. Dogs learn most of their socialization skills between the ages of three weeks and three months. Most reputable breeders won't allow pups to be separated from their mothers before they're eight weeks old; in many cases they leave the litter much younger, so they never receive the lessons in Puppyhood 101 they will need to become well-adjusted adult dogs. It's a vital phase that is too often missed when pups are sold too soon to turn a quick profit. Fortunately, this practice has become a thing of the past, at least in the state in California. In 2005, Governor Schwarzenegger signed Senate Bill 914, which prohibits the sale of puppies younger than eight weeks old, with punishment for violations of up to $1,000 and a year in jail.

My Daisy was the product of a puppy mill, one of those mass breeders of ill repute that abound in the Midwest. I didn't know that until years later, when her former owner got in contact with me and told me about her checkered past. But I already knew all I needed to know about Daisy's life before she reached me. How did I know that? First, there is the fact she's a total nut job—a lovable nut job, but a nut job, nevertheless. Perhaps even more indicative, when I adopted her, I could see immediately that she had acquired virtually no social skills in her first year of life. She seemed to have no familiarity with the basic commands any one-year-old dog should know, such as "Sit," "Stay," "Down," and "Give me back my hand, you crazy dog!"

Daisy is an abnormal dog, primarily because of her lack of socialization at the most crucial time in her development. Most of her training occurred after she came to live in our home, but by the time a dog is a year old, it's much harder to teach him what he needs to know about being a good family pet. The time to learn those things is when a pup is still with its mother.

Irresponsible breeders just keep creating more Crazy Daisies for unsuspecting owners to adopt. Daisy was one of the lucky ones whose owner didn't give up on her (although I confess there were times it crossed my mind), as many owners might have. If I had, she might have ended up in a pound or shelter, where she may not have found another home. She certainly would have flunked the social tests dogs are routinely given before they can be adopted out to a prospective family. Puppies should stay with their mothers and siblings until they are at least eight weeks old—with some breeds, the ideal minimum age is twelve weeks or even longer—during which time they will learn all there is to know about being a good dog.

> "Whoever said you can't buy happiness forgot about puppies."—Gene Hill

The initial nine weeks, known as primary socialization, is an important discovery phase for a pup. During this stage of development, he learns basic skills like self-control, communication, and how to find his place in the pack hierarchy. This primary socialization period is one of intense learning that allows a puppy to more easily adjust to his environment, both now and in his future home.

At three weeks, the pup begins what is known as the attraction phase, in which he experiences surroundings with no apprehension. This is when you put away your good shoes and anything else chewable. At this time, a pup should come in contact with species unlike his own. Pups should be handled (gently, of course) so they will become accustomed to humans and identify them as friends throughout life. Just like a human toddler, this is the time at which a dog's capacity for learning is peak and memorization is best. Learning is easy and breezy—he won't ever forget the lessons he's taught at this time. During this phase, a puppy learns first of all that he's a dog. He must live with a species unlike his own—that is, the human species. He must communicate, and biting is a no-no!

The period of attraction begins to wane when the pup reaches the age of nine weeks, when it enters an aversion phase, also known as the fear period. Although he retains the social attachments he has formed up to this point, he won't be as willing to form new ones. This is the time when the puppy learns that not everything in the world is his friend—there are things out there that can hurt him. He will tend to be fearful of new things or strange people and might, without your help and guidance, automatically perceive them as enemies.

At this point, the social imprinting on a dog's brain can easily become permanent, so it's up to the dog's owners to increase the number of contacts and experiences that the dog receives during this phase. Whether or not the dog gets these things can depend a lot on how well socialized the dog's owner is. Sometimes an owner is overly protective of a young dog and won't allow other dogs or people to come in close contact with him. This teaches the dog to become standoffish or even aggressive around other dogs or people. If the dog's owner allows this antisocial behavior, or worse, rewards it, that dog will become a problem pet and may end up being surrendered once the cuteness of puppyhood is past. We've probably all encountered the snappish lapdog whose owner thinks is utterly adorable, even when he's baring teeth at you. That behavior wouldn't be tolerated in a larger breed of dog, which could do some serious damage with those formidable choppers.

A pup that doesn't learn not to mouth his owner or how to control his instinct to bite will grow to think it is acceptable to bite a human. That play-biting behavior that was cute in a six-week-old pup won't be so cute when the grown dog bites a child and becomes a four-legged lawsuit.

It becomes more difficult for a dog to learn these bow-wow basics at later stages of development. At this stage of your dog's development, it's good to take him along with you to dog parks, children's playgrounds, or other public places where he is sure to come in contact with other animals and humans. This way, he learns not to fear new associations or experiences that he might encounter later on as an adult. After the pup passes through this primary socialization period, he enters what is called the juvenile period. During this time, he's very much like a four-legged teenager.

Even if, as in Daisy's case, the primary socialization period has been lacking, there's still hope with another type of socialization called secondary socialization. Still, teaching a dog what it needs to know at this stage is more difficult and less effective than getting a dog off on the right paw to begin with.

Chapter 2

Intelligent Dog Design

A dog is not almost human, and I know of no greater insult to the canine race than to describe it as such.—John Holmes

When it comes to the (dare I say the word?) evolution of the canine species, proponents of the theory of intelligent design will be thrilled to know that man, in his infinite wisdom, has had a creative hand in the development of the nearly 500 recognizable breeds of modern-day dog.

Doggie DNA: The History Behind Man's Best Friend

Thanks to the studies of doggie DNA, it has been proven that dogs evolved from wolves, but genetics cannot tell us for certain when our association with dogs first began. One possibility is that dogs were created at the same time as humans because God knew that man was going to need somebody intelligent (besides Eve) who would put up with him. With a dog around, it was no problem if Adam ate like a slob and tossed ribs in Eve's flowerbed. Whenever it was that our bond was formed with the dog, everyone agrees that it was a doggone long time ago.

According to experts, the prehistoric ancestors of our present-day canines first appeared on earth 10 million years ago.

While man and dog may have first interacted as far back as 500,000 years ago, most agree that it was about 10,000 years ago that humans and dogs figured out there was something to be gained by forming a partnership. We probably first used wolves to help us hunt for our food when we discovered they were a lot better at bagging the wily wooly mammoth than we were. As to how much longer it took for the two species to progress from "Ugh, ugh, woof, woof" to "Off the bed!" or "Go lie down!" one can only guess.

"Ever consider what they must think of us? I mean, here we come back from a grocery store with the most amazing haul—chicken, pork, half a cow. They must think we're the greatest hunters on earth!"—Anne Tyler

Dogs have been serving us faithfully since we were tagging graffiti on cave walls. As humans progressed from cave dwellers to hut dwellers and formed territorial tribes, dogs provided all kinds of useful services for us. They warned us of intruders, guarded our property, herded cattle, and tended the flocks. In return, they were well fed and cared for. It was a mutually beneficial alliance for both species then, and it still is.

From Half-Breed to Purebred

While all members of Canis familiaris share common characteristics, it didn't take humans long to realize that these creatures they'd succeeded in domesticating could be useful for many different tasks. Beyond that, people discovered that by breeding dogs with certain physical or behavioral traits, they could design a dog that was well suited to a particular task.

> That's the thing with people—we're always thinking about ourselves and our needs. That rule applies to the way we went about developing the various dog breeds, some of which would **never** have occurred in nature! Never mind the fact that a dog bred to run around on short little legs might trip on his long floppy ears. Or the dog with long curly bangs might need a haircut every so often to be able to see where he was going. As long as we're happy, that's all that matters!

Over time, the practice of purposeful breeding has given us the broad variety of purebred dogs we see today. Every breed of dogs has not only physical but behavioral differences. Those are important considerations when selecting a dog. We've said it before, but it's worth saying again: When you choose a breed of dog, you should pay attention to those breeds with traits that are suitable for your lifestyle. If you are a sedentary person, you would probably be better off choosing a breed that is not too energetic. It would therefore be wise to avoid breeds that require a great deal of exercise, such as herding breeds like Border collies and Australian cattle dogs, or other workaholic breeds that must be working in order to be happy. The same is true of hunting breeds, whose instincts are not very far removed from those of their wild ancestors. These dogs still want to be out in the field bagging game with their human hunting companions, although these days the game is a bit smaller than a mastodon.

Beyond Book Smart

Up . . . up . . . and away! Forget **mental** leaps and bounds . . . a greyhound named Cinderella May A Holly Gray cleared the highest jump ever recorded when she jumped 66 inches in 2003.

Dogs are still our closest allies in the animal kingdom, and they still serve us in myriad ways. Some just happen to be sillier than others. You've probably noticed that Britney

Spears doesn't carry a Komondor around in that designer doggie bag of hers (not that she could, considering the breed's formidable size). Britney may be a high-maintenance gal, but she has a low-maintenance dog—tiny, totable Bit Bit. A Chihuahua is the perfect toy breed for someone like new mommy Britney who is always jet-setting around the globe. Adored pet or fashion accessory? You decide.

Of course, whether they're large or small, energetic or just plain lazy, dogs are all equal in their capacity to give us companionship and affection. Without a doubt, that is the greatest benefit we've ever received from our relationship with the dog.

Dogs That Might Not Make the Grade

New breeds of dogs are constantly being developed from crossing one breed with another, although it may take years for them to be accepted by the American Kennel Club (AKC), if they ever are. Somehow, I doubt that a Chiweenie, a cross between a Chihuahua and a dachshund will make the cut, but you never know. Some mixes have found their way into the mainstream, such as these:

- Labradoodle (Labrador retriever crossed with poodle)
- Schnoodle (schnauzer crossed with poodle)

- Peekapoo (Pekingese crossed with poodle)
- Cockapoo (cocker spaniel crossed with poodle)
- Maltipoo (Maltese crossed with poodle)
- Spoodle (Spitz crossed with poodle)
- Yorkiepoo (Yorkshire terrier crossed with poodle)
- Puggle (pug crossed with poodle)

All of the breeds now recognized by the AKC were the result of a little doggie DNA dabbling over the years. Here are some other curious canine crosses (from *No Bones About It: New Canine Breeds* by Deborah Schmoll) that may eventually find a class of their own in the AKC registry (but probably not):

- Collie + Lhasa apso =
  Collapso, a dog that folds up for easy transport.

- Pointer + setter =
  Poinsetter, a dog that likes to decorate your home for the holidays.

- Cocker spaniel + poodle + Doberman =
  Cocka-poodle-doo, a dog that's wise and early to rise.

- Pekingese + Lhasa apso =
  Peekasso, an artistic dog.

- Poodle + Great Pyrenees =
  Poopyree, a dog that smells good.

- Smooth fox terrier + chow chow =
  Smooch, a dog that's always affectionate.

- Bull terrier + shi tzu =
  Never mind!

We might never see these new breeds, but science is forging ahead to create some designer dogs. They may not be as silly, but the concept itself strikes some dog lovers as more than a little scary.

Beyond Book Smart

That agility training really took off on December 4, 1999, for Jazz, a border collie, when he weaved between sixty poles in a mere 12.98 seconds in George, South Africa.

Besides the obvious differences in traits that have been bred into dogs, there are also variations in the personalities of individual dogs. Like people, each dog is unique in its makeup. Every dog is born with a different genetic imprint. Its experiences and upbringing are different. It looks different from other dogs. No two dogs behave the same, either. If you were to clone your dog, the new version would not be the same as the dog you had before.

Some would argue the point, but in my estimation the thing that would be missing from the canine copy would be the original animal's essence, or soul, if you will. That's another reason not to spend $50,000 on doggie duplication. Besides, should humans really be playing "doG"?

Breeding for Personality

There's some weird wagging science going on these days, as you'll discover if you've scanned the "Pets for Sale" section in your local newspaper or recently visited a pet shop that is still in the business of selling dogs.

Whereas you might once have seen only purebreds for sale, nowadays you are more likely to see unique blends of those breeds. Through the practice of mix-and-match genetics, breeders are striving to design a better dog. Many people suffer from allergies and can't have a dog that sheds. Since most designer mixes are a blend of poodle and some other breed, that means they are hypoallergenic. An added bonus is that your designer mutt won't shed white fur all over your black gabardine slacks or black fur on your white suede sofa. Poodles are also longer-lived than some other breeds as well as more intelligent, agile, and energetic.

"I wonder if other dogs think poodles are members of a weird religious cult."—Rita Rudner

Not everyone may want a poodle, though, so breeders are creating plenty of other mixes by selecting the best traits of the purebreds and passing those on to the crossbred dogs. Every breed has desirable qualities and also undesirable ones. For instance, while a golden retriever may be gentler and easier to train than some breeds, goldens also have a tendency to have hip dysplasia. Cross a golden with a poodle, and you get a calmer Goldendoodle that has sound hips, doesn't shed, and doesn't have the skin problems that poodles and other non-shedding breeds often have.

Of course, these supermutts don't come cheap. Often, you'll pay as much or more than you would for a purebred dog. It's easy to worry that this phenomenon may generate more of the same problems that other canine crazes have done in the past. Puppy mills and the misery they generate thrive on the materialistic demand created by the desire to be the first one on your block to own a Labradoodle or Yorkipoo or other designer dog.

"Among God's creatures, two, the dog and the guitar, have taken all the sizes and all the shapes, in order not to be separated from the man."—Andrés Segovia

In Search of a Breeder

When looking for a breeder, whether for the intent of purchasing a purebred or one of the designer dogs so many people are turning to, it is important to find one you can trust. Irresponsible breeders are no more likely to take care in breeding a canine cross than in a purebred dog. It's just as important to take care when choosing a crossbred dog as it is with any dog. You'll be living with this dog for the duration of his life, right? That means you want to be sure that the dog has no genetic health problems that could come back to bite you in the wallet later or set you up for a big heartache.

> If you can't decide between a shepherd, a setter, or a poodle, get them all—adopt a mutt!—ASPCA

As it is with all types of pets, it's wise to shop carefully for a purebred dog. When choosing a pup, use your head as well as your heart. Research the breeder's track record, and thoroughly check the genetic history of your prospective dog's parents. If the pup is truly the product of purebred parents, you should be able to trace their pedigree back several generations. Just as there are plenty of reputable breeders, there are also dishonest breeders who would cross a poodle with a platypus (a platypoodle has a nice ring, don't you think?) if they could make big bucks from an unsuspecting buyer. They also don't care if they breed puppies from a closely related sire and dam. Too often the pups are the product of unions between siblings or mothers and sons, which can result in all kinds of genetic anomalies and resulting health problems.

Here are some things to watch for when choosing any puppy from a breeder:

- Is the kennel sanitary and well ventilated?
- Do the animals look clean, healthy, and well cared for?
- Do the pups seem well socialized to humans?
- Are the pups at least eight weeks old? (If not, that means they have not been with their mother long enough to be well socialized.)
- Are both the pup's parents on the premises so you can meet them?
- Did the breeder interview you before even talking to you about a pup? This can be off-putting to some prospective buyers, but it's a sign that the breeder really cares about the kind of home the pup will be raised in. The best breeders may even have a waiting list for their pups because they always ensure the best possible pups go to the best possible homes.

- Is the breeder willing to offer a money-back guarantee if for some reason things do not work out with you and your new ball of fluff or if there are health problems?
- Can the breeder verify the ancestry of the pups?

While some people feel that AKC registration papers aren't good for much more than for lining the bottom of a whelping box, they do serve to trace the parentage of a purebred litter. The AKC mandates that three generations is the minimum for a dog to be qualified for registration as a purebred. What those papers do not necessarily verify is the soundness of the dogs in those generations. They don't provide any assurance of the absence of inherited health problems that you might be purchasing along with that AKC-papered pup. Testing for genetic defects before selling a litter of puppies is the responsibility of the individual breeder.

> "If a dog doesn't put you first, where are you both? In what relation? A dog needs God. It lives by your glances, your wishes. It even shares your humour. This happens about the fifth year. If it doesn't happen you are only keeping an animal."—Enid Bagnold

Is Your Dog Smart Enough to Breed?

If that Pomeranian that's romancing your leg could talk, he'd probably tell you that he's in the mood for love and has what it takes to make more cute little Pomeranians. But does he? Just because he's a purebred doesn't mean he's fit to father any puppies.

Many people seem to think that just because they have a purebred dog, it automatically means the dog should be bred. More often than not, that papered purebred may be harboring some serious genetic defects in his well-watered family tree. These problems may not always be apparent, but they can still be passed down to future generations of pups and to unwary owners. There's an important little factor called DNA that should be investigated before any dog is allowed to breed with another dog.

"Money will buy a pretty good dog, but it won't buy the wag of his tail."—Josh Billings

Fortunately, just as there are tests for people to uncover any potential gene anomalies, there are also tests to determine whether your dog is carrying any genetic defects. It's as simple as drawing a sample of blood to pinpoint some of the more common hereditary disorders in dogs. These include hip and elbow dysplasia, epilepsy, von Willebrand's disease (a bleeding disorder similar to hemophilia in humans), and allergies, to name only a few. These tests are relatively inexpensive, especially when you consider what it can cost to treat these disorders later on.

Beyond Book Smart

Gibson, a great Dane, can really move to the head of the class—because he stands heads above the rest at over 42 inches tall!

Clearly, irresponsible breeding practices are at the root of these and other hereditary diseases in purebreds. "It's a people problem," says Dr. Jean Dodds, founder and president of Hemopet, a nonprofit animal blood bank and greyhound rescue program in Southern California. "There's nothing wrong with controlled inbreeding and line breeding; it fixes type, as long as you know what you're doing. The problem is being honest about what you have and don't have." She goes on to say, "For every fifty breeders that are trying to do the right thing, you've got a handful that are ruining it for everybody else."

Roscoe Jones, the yellow Lab, loved to eat anything and everything out of the bathroom garbage can. His owner, Jacky, purchased a small can with a lid. Roscoe followed her into the bathroom and noticed that as she stepped on the can's pedal, the lid lifted. Triumphantly, Jacky threw her tissue into the can and watched the lid shut tight. Roscoe immediately walked over to the can, pressed the pedal to lift the lid, retrieved the tissue, and ate it with great relish.

If you haven't tested the soundness of your dog's genetic makeup, or you already know that he has health problems that could be passed on, do the responsible thing by spaying or neutering your pet. The shelters are already full of unwanted pets, and besides, it's not nice to litter.

Chapter 3

Why Hounds Hunt and Terriers Tear

Englishman Jimmy Dawson's wire-haired terrier, Queenie, was highly intelligent, as is typical of the breed. She was also very observant. Jimmy kept his dentures beside his bed at night. One morning, he couldn't find them anywhere. Frustrated, he went into the kitchen to ask his wife if she'd seen them, when Queenie padded into the kitchen.

"Will ye have a look at the dog, Jim." Jimmy looked down at Queenie. With Jimmy's choppers held fast in her own, Queenie shot him a toothy grin.

The old Southern expression, "That dog won't hunt," means "That idea won't work." And by that, I mean it jest ain't fittin' to apply the same set of intelligence standards to all dog breeds. Just as there are difference races of human beings, all with varying physical traits, there are different breeds of dogs, and every breed was bred for a special purpose. Herding dogs like Border collies and Australian cattle dogs herd. Sight hounds like greyhounds and salukis respond to visual stimuli. Scent hounds like bloodhounds, basset hounds, and blue tick hounds sniff out their cotton-tailed quarry. Rat terriers rat. Terriers tear.

> The word "terrier" comes from the Latin word terra, which means earth, and the name refers to the fact that terriers dig in the earth to seek out their underground prey. They also love to dig up the begonias you just planted in your garden.

In purebreds and crossbreds alike, certain traits that were bred into the original breed will be evident to a greater or lesser degree in the dog. No matter what variety of retriever you own, your dog will have a propensity for retrieving something, even if it's only a tennis ball. Hound dogs will track whatever they were bred to track: hare, deer, elk, fox, otter, or in the case of my Bubba Gump, cats and biscuits. And although Miss Manners might heartily disapprove of such rude behavior, a pointer will always tend to point.

The AKC classifies dog breeds into several groups: sporting, hound, working, terrier, toy, nonsporting, herding, and miscellaneous. The breakdown of the groups can be found on the following pages. Somewhere on this long list of breeds is a dog to suit every pet preference. You may not even be familiar with some of the exotic-sounding breeds, and this

What's What: The AKC Breakdown of Dogs

The Sporting Group

American water spaniel
Brittany spaniel
Chesapeake Bay retriever
Clumber spaniel
Cocker spaniel
Curly-coated retriever
English cocker spaniel
English setter
English springer spaniel
Field spaniel
Flat-coated retriever
German shorthaired pointer
German wirehaired pointer
Golden retriever
Gordon setter
Irish setter
Irish water spaniel
Labrador retriever
Nova Scotia duck tolling retriever
Pointer
Spinone Italiano
(not a flavor of ice cream)
Sussex spaniel

Vizla
Weimaraner
Welsh springer spaniel
Wirehaired pointing griffon

The Hound Group

Afghan hound
American foxhound
Basenji
Basset hound
Beagle
Black-and-tan coonhound
Bloodhound
Borzoi
Dachshund
English foxhound
Greyhound
Harrier
Ibizan hound
Irish wolfhound
Norwegian elkhound
Otterhound
Petit basset griffon vendeen
Pharoah hound
Rhodesian ridgeback
Saluki

Scottish deerhound
Whippet

The Working Group

Akita
Alaskan malamute
Anatolian shepherd
Bernese mountain dog
Black Russian terrier
Boxer
Bullmastiff
Doberman pinscher
German pinscher
Giant schnauzer
Great Dane
Great Pyrenees
Greater Swiss mountain dog
Komondor
Kuvasz
Mastiff
Neapolitan mastiff
Newfoundland
Portuguese water dog
Rottweiler
Saint Bernard
Samoyed

Siberian husky
Standard schnauzer

The Terrier Group
Airedale terrier
American Staffordshire terrier
Australian terrier
Bedlington terrier
Border terrier
Bull terrier
Cairn terrier
Dandie Dinmont terrier
Glen of Imeal terrier
Irish terrier
Kerry blue terrier
Lakeland terrier
Manchester terrier
Miniature bull terrier
Miniature schnauzer
Norfolk terrier
Norwich terrier
Parson Russell terrier
(formerly known as the Jack Russell)
Scottish terrier
Sealyham terrier
Skye terrier

Smooth fox terrier
Soft-coated wheaten terrier
Staffordshire bull terrier
Welsh terrier
West Highland white
Wire fox terrier

The Toy Group

Affenpinscher
Brussels griffon
Cavalier King Charles spaniel
Chihuahua
Chinese crested
English toy spaniel
Havanese
Italian greyhound
Japanese chin
Maltese
Manchester terrier
Miniature pinscher
Papillon
Pekingese
Pomeranian
Poodle
Pug
Shih tzu

Silky terrier
Tiny fox terrier
Yorkshire terrier

Nonsporting Breeds
American Eskimo dog
Bichon frise
Boston terrier
Bulldog
Chinese shar-pei
Chow chow
Dalmatian
Finnish spitz
French bulldog
Keeshond
Lhasa apso
Löwchen
Poodle
Schipperke
Shiba inu
Tibetan spaniel
Tibetan terrier

The Herding Group
Australian cattle dog
Australian shepherd

Bearded collie
Belgian Malinois
Belgian sheepdog
Belgian Tervuren
Border collie
Bouvier des Flandres
Briard
Canaan dog
Cardigan Welsh corgi
Collie
German shepherd dog
Old English sheepdog
Pembroke Welsh corgi
Polish lowland sheepdog
Puli
Shetland sheepdog

The Miscellaneous Class

Beauceron
Plott
Redbone coonhound
Swedish vallhund
Tibetan mastiff

is just the tip of the dewclaw. There are many other breeds in existence. The dog breeds listed in each group share basic qualities and characteristics, but the range of intelligence and trainability varies from breed to breed within a group.

Group I—Sporting Breeds

Comprising the group of sporting breeds are pointers, setters, and spaniels. When Elmer Fudd is hunting in a field or hunkered down in a duck blind, he's going to want a dog that can alert him to the presence of Bugs Bunny or that will bound eagerly into water to retrieve Daffy Duck. Probably the most popular of these are the Labrador retriever, the golden retriever, and the weimaraner. The German shorthaired pointer is a preferred all-purpose dog among hunters. Highly intelligent, this breed is descended from breeds such as the English foxhound, with its exceptional scenting and tracking abilities. While a bird in the paw may be worth two in the bush, these dogs can sometimes be too focused on the task at hand . . . kind of like an engineer or computer techie.

Group 2—Hounds

Hounds are hunters of a different type. They were bred to follow prey by scent, like the bloodhound, coonhound, or basset hound, or by sight, like greyhounds, whippets, and salukis. "Zoom, zoom, zoom" is not just a good slogan for a sports car, it works just as well for the speedy sight hounds, which can spot prey from a great distance and close in on the unsuspecting creature in the flick of a whisker. The Afghan hound is a coursing hound that can outdistance riders on horseback. This breed from Afghanistan prefers to hunt singly rather than in packs, the way most hounds tend to work, which accounts for its rather aloof nature. So not only does the gorgeous blonde Afghan look a bit like Greta Garbo, if she could talk, she'd probably say, "I vant to be left alone."

Beyond Book Smart

Olive Oyl, a Russian wolfhound, is the envy of every schoolyard. She holds the world record for most jump rope skips done by a dog in one minute, at sixty-three skips.

Scent hounds, on the other hand, are slow, methodical trackers with a highly developed sense of smell, the keenest of any of the breeds. In no particular hurry to get where they're going but determined to do so, these dogs were bred to follow the scent of game for miles, through all kinds of terrain. Of course, the longer-legged bloodhound can cover those miles a lot quicker than his short-legged cousin, the basset hound. Once you can get the blasé basset off the sofa, either breed eventually will get from point A to point B—they just won't get there at the same time. Some hounds have even tracked a scent across bodies of water, though the shorter-legged breeds would have to do their dogpaddling in a canoe because they aren't very good swimmers. Don't expect a dachshund or a basset hound to tread water as well as a pointer or retriever. Even though they also have a respectable hunting heritage, they'll come up short every time.

"So many get reformed through religion. I got reformed through dogs."—Lina Basquette

Group 3—Working Breeds

Heigh, ho, heigh, ho, it's off to work the working breeds go. The career choices for these breeds are as varied as the breeds themselves. Some are fierce protectors, like the mastiff, pinscher, or rottweiler. Some are trekkers across frozen tundra, like the Alaskan malamute and Siberian husky, used in the Iditarod cross-country dogsled races. But all working breeds have one thing in common: They are dedicated workaholics that will never expect a paid two-week vacation in the Bahamas.

Group 4—Terrier Breeds

If you have a little trouble getting the attention of a terrier, it's because he's much too busy doing more important things, like digging the hundredth hole in your garden. What else would you expect a dog whose name means "earth" to do? Ranging from the larger Airedale to the diminutive Dandy Dinmont and West Highland white, these scrappy, agile fellows with keen scenting ability were actually bred to pursue small prey in their underground lairs. In the absence of a rabbit or a badger to root out, the terrier will devise all kinds of clever games to entertain himself, such as bounding from sofa to chair to floor, chasing his tail, or playing leap dog with himself. You won't find a more engaging or playful companion than the terrier, but like a hyperactive child, if left to his own devices for long he can be a holy terrier.

> "Things that upset a terrier may pass virtually unnoticed by a Great Dane."—Smiley Blanton

Group 5—Toy Breeds

These are the dogs that people love to spoil. They are the lapdogs of celebrities and European royalty. From Sharon Osborne to Paris Hilton, toy breeds have found their place among the stars. Walk through any art museum, and you'll see paintings of bejeweled lords and ladies of long ago doting on their cavalier King Charles spaniels, pekes, or pugs. Adorable, affectionate, and gentle, these dogs were bred for the good life. They belong indoors, close beside their besotted owners, who sometimes have a tendency to spoil them a bit too much. Because we have bred them to be our toys, these delicate breeds do require much more of our attention and care than some other breeds might, but of course all dogs love our attention.

Group 6—Nonsporting Breeds

In this class you'll find the merry, hairy, and downright scary. They aren't included with others because they are unique in their appearance, heritage, or odd characteristics. And some have rather spotty backgrounds. The dalmatian falls in this group. This intelligent breed has been running alongside coaches since Egyptian times. While not all the dogs in this group are bred for their beauty, like the Chinese shar-pei, there are some notorious brainy breeds in this pack. The poodle, for instance, originated as a water dog. Poodles have long been revered for their intelligence and trainability, which is why they are often seen in dog acts performing tricks. But oh, what to do with all that hair!

Group 7—Herding Breeds

You have no doubt heard about some of the notable herders in this group of dogs. Of course, everyone knows about Lassie, the famously intelligent collie of television fame or the Bouvier des Flanders of the well-loved story, "A Dog of Flanders." There's also Hank the Cattle Dog, a popular children's book character. Another literary canine, the Belgian Malinois, is featured in thriller novels written by John Lescroart.

Beyond Book Smart

You'd be surprised what your dog is capable of when presented with the world's largest dog biscuit, created in 1999 by the People's Company Bakery in Minneapolis, Minnesota. At over 7 feet long and almost 2 feet wide, your dog would have to do a lot of sitting and staying to deserve that treat!

Some herding dogs serve double duty, like the German shepherd and the Malinois, which are also used in search-and-rescue work. This crossover tendency is true of many breeds. Dogs are not only lovable and devoted but also versatile. Even though the various dog breeds were bred for specific tasks, some dogs often possess other unique talents that can also serve man. You never know, you could have a pup prodigy on your hands. If only there were a canine college aptitude test!

Group 8—Miscellaneous Class

This canine classification "miscellaneous" might make these dogs seem like afterthoughts, but they actually do belong in a class all their own. The Miscellaneous Class is the AKC's way of acknowledging new breeds and recognizing them by allowing them to compete for championship points in conformation competition and by maintaining an official stud book.

That does not mean that all breeds in the Miscellaneous Class are new in the sense that they've just been developed. The Swedish vallhund, for instance, is an ancient breed that has been known since Viking times for its cattle-herding abilities. The beauceron is another old herding breed that is barely known outside its native France. Unlike most other purebreds, the beauceron was developed entirely from native French stock, with no crosses to foreign lines, giving this dog a pedigree that some claim can be traced back to Renaissance times.

Beyond Book Smart

Hot dog! This dog may not be smart, but he can sure fill you up! The longest hot dog ever made was over 34 feet long!

The AKC admits new members to the Miscellaneous Class when it is demonstrated that there is strong U.S. interest in breeding, working, and/or showing dogs of that breed. Dogs in this class can compete in AKC events such as agility, obedience, and tracking, and they are eligible for conformation competition at the Junior Showmanship level only. When the AKC is satisfied that interest in the breed is strong ***and*** growing nationwide, breeds in this class may be admitted to the appropriate group (Terrier, Herding, Nonsporting, and so on).

How many dogs does it take to change a light bulb?

Golden retriever: The sun is shining, the day is young, we've got our whole lives ahead of us, and you're inside worrying about a stupid burned-out light bulb?

Dachshund: You know I can't reach that ridiculous lamp!

Rottweiler: Make me.

Shih tzu: Puh-leeze, dah-ling. Let the servants . . .

Lab: Oh, me, me!!! Pleeeeeeze let me change the light bulb! Can I? Can I? Huh? Huh? Can I?

Parson Russell Terrier: I'll just pop it in while I'm bouncing off the walls and furniture.

Mastiff: Mastiffs are *not* afraid of the dark.

Chihuahua: Yo quiero Taco Bulb.

Pointer: I see it, there it is, there it is, right there . . .

Greyhound: It isn't moving. Who cares?

Australian shepherd: First I'll put all the light bulbs in a little circle . . .

Hound: ZZZZZZZZZZZZZzzzzzzzzzzzz

The Ratings Game

People are obsessed with scores. From sports to *American Idol* competitors to Internet polls, it's all about ratings. It's no different with dogs—we naturally want to know where our dog rates on the intelligence sale. Dogs are like people. Some are geniuses and some are average. Others rank pretty doggone low, but there are some important government officials who would rank pretty much the same.

There are categories of dogs that consist of bullheaded, difficult to train breeds of dogs, but maybe that's because they have their own ideas about how (...and when...and why) to do things. Perhaps that's the reason some don't respond to commands as quickly as we would like them to. Stubbornness doesn't necessarily indicate low intelligence, although that is not the case with some people. The Scottish terrier has a good excuse for his stubborn streak. He spends too much time with presidents. The basset hound is also known for its stubborn streak. It's a fact that hound dogs have a mind of their own, but being a deliberate and independent thinker has its pluses. Regardless of where dogs rank on the intelligence tests people make up for them, every dog ranks number one in the ability to give you the most important qualities a dog has to offer his owner: love and companionship.

There are many levels of dog intelligence. If your dog obeys a command even before you get the words out of your mouth, or if you catch him writing quantum physics equations on his doghouse walls, your dog clearly has a beautiful mind. The lively Welsh Corgi is a favorite among Great Britain's royal family, but that should constitute no reflection on the dog's intelligence. It's sometimes said that the Airedale can be kind of an Airehead, and the Dalmation might need some regular coaching to keep him fulfilled in his work.

Having a dog that can outthink you can have its drawbacks, though. If you look into the bright, intelligent eyes of a Border collie, you'll be convinced he's plotting a canine coup. You may have seen the same cunning in the eyes of your own dog, even though he might never rank on the lists of smartest dogs devised by experts in dog intelligence. But even dogs that don't rate at genius level on dog tests should be awarded gold chokers when they retire and get their AARF (American Association of Retired Fidoes) cards just for putting up with us exasperating humans.

Stanely Coren rated dogs according to their ability to understand and respond to commands. But what if the tables were turned and it was your dog rating *you* on your intelligence and trainability? Would a dog's estimation of your brainpower put you at the top of the IQ scale—or at the bottom of the dog pile of owner intelligence? If your dog is anything like most people's dogs, he probably has you better trained to carry out his commands than you'll ever have him trained to carry out yours. Most of us have become slaves to our dogs, which isn't always a good thing, at least from some people's perspective. When your dog barks his commands to open the door or deliver him a biscuit, you no doubt respond, but probably not as quickly as he would like.

If dogs were to come up with a rating system that helped them to choose the most intelligent and trainable owner, it might look quite a bit different than you'd expect. If you'll excuse the expression in a book about dogs, here are the *categories* that might be used to determine human intelligence and trainability, according to your dog's point of view.

The number of barks required for you to respond to your canine's command determines your rank on the dog's intelligence scale for owners:

- **Lassie level:** 5 barks or fewer
- **Eddie level:** 6 to 15 barks
- **Spot level:** 16 to 25 barks
- **Satchel level:** 26 to 40 barks
- **Quincy level:** 41 to 80 barks
- **More than 80 barks:** Fuggedaboudit! Your human is hopelessly untrainable.

Your Canine's List of Commands

- Open the door to let me out, even though I have a dog door of my own.
- Open the door to let me back in.

- Give me the doggy bag of leftovers after dining out.
- Throw a ball or Frisbee without tiring.
- Leave the Sunday roast on the table or counter—within my reach.
- Let me chase the cat.
- Let me chase squirrels and anything else that dares step in *my* yard.
- Offer frequent and copious amounts of treats.
- Feed me scrumptious tidbits from the table.
- Let me pee on the neighbor's plants if I want to.
- Let me dig in the garden to my heart's content.
- Let me bring my buried treasures into the house.
- Let me roll in whatever I like, no matter how disgusting.
- Don't insist on bathing me after I've had a good stinky roll.
- Don't yell at me when I have an accident in the house.
- Leave the toilet lid up so I can drink from the toilet.
- (Applies to breeds with pendulous lips, like bloodhounds, mastiffs, or basset hounds): Let me slobber and sling drool from one end of the house to the other.
- (Applies to breeds that shed): Let me get white fur all over black clothes/furniture/carpet or black fur on white clothes/furniture/carpet.
- Take me to work with you.
- Give me a present at Christmas, on my birthday, and other howlidays.

- Make sure I always have my current address and phone number on my collar—even if I can't read it, it's nice to know it's there.
- Give me my own cell phone and answering service, just in case I'm visiting with the neighbor's poodle (wink, wink).
- Never say the "N" word (neuter) within earshot of my girlfriend, Fifi, even though we both know there's no Viagra for dogs.
- Don't dress me in silly dog costumes to embarrass me in front of your buddies.
- Take me to the dog bakery and let me choose my own pup pastries.
- Bake homemade dog biscuits for me.
- Serve only the best, most expensive canine cuisine.
- Never serve me the same meal two days in row.
- Take me for a daily walk.
- Take me for more than one daily walk.
- Take me to the dog park at least once a week.
- Take me to the dog park more than once a week.
- Take me along on the family vacation. I'm part of the family, aren't I?
- Buy me a new toy once a month.
- Buy me a new toy every week.
- Let me sleep in the house.
- Let me sleep on the couch.
- Let me sleep on the bed.

- Give me a daily brushing.
- Give me a nose-to-tail body massage.
- Let me stop and sniff the roses for as long as I want when out on walks. Also, let me pee on them.
- Don't use a choke chain to lead me.
- Let me lead you, *not* the other way around.
- Don't holler at me when I scoot on the carpet.
- Leave the trash bin full of delicacies like maggoty meat (extra protein, Yippee!) and dirty diapers.
- Allow me to sample hors d'oevres from the cat's litter box.
- Let me give you a kiss after I eat the catbox crispies and the gourmet garbage.

So how would you do on your dog's intelligence test, if he gave you one? Where do you think you'd rank on the list? Would you be a Lassie or a Quincy? The truth is, dogs rate quite a lot higher in many areas of intelligence than we do, and those areas have more to do with measuring the qualities of the heart rather than the brain.

What breed of dog are you?

It's been said that some people look like their dogs, which can be either a compliment or an insult, depending on the breed you are referring to. For example, if you are the owner of a bulldog or a Mexican hairless and someone says you look like your dog, you may just want to bite that person on the leg. Then again, if you think your dog is beautiful, as most dog lovers do, you might not be insulted in the least to be compared to your dog. Even the woman who owned Sam, who was voted the world's ugliest dog, didn't think he was ugly. Although, I'm not sure she would like to have been told she looked like him.

People not only have a tendency to share some physical characteristics of the dogs they own, but they also seem to share other characteristics of their favorite breed. When driving in your car, you may at some time have found yourself following in traffic behind a couple of gorgeous gals zipping along in a flashy red convertible. One of them is a brunette, but you're most attracted to the blonde in the passenger seat. Their long, wavy tresses ruffle in the wind. You're intrigued, allured. At the next stoplight, you decide to make your move and get to know the blonde a little better. You pull up next to their car for a closer view of this platinum-haired goddess. Quickly, you check your look in your rearview mirror, just in case there's any spinach from lunch left dangling between your teeth. You steer your battered family sedan alongside the passenger's side of the sports car with the two hotties, roll down the window with the peanut butter and jelly handprints on it, and get ready to deliver your best flirty smile. Your heart beats a little faster as the blonde turns to look at you. It's only then that you realize you've been tailing a woman and her Afghan hound.

In the latest version of *The Shaggy Dog*, Tim Allen is transformed into a dog after being bitten by one, sort of an American werewolf in Hollywood. If the scenario in this movie could actually happen and you somehow were morphed into a mutt, what kind of dog might you become? What breed do you think you share the most similar characteristics with? If you're not exactly sure, then consider the following:

Are you a highly intelligent person who can outthink the best of them and who graduated at the top of your class at an Ivy League college? Perhaps you are a research scientist, a mathematician, or a college professor. If you are, then you might be the smartest dog, a Border collie. If you're more brawn than brain, though, you are probably a bulldog.

Are you an athlete who runs marathons? Are you a bi-athlete? A tri-athlete? Maybe you are even an Olympic competitor. Do you have a lithe, graceful body with long legs? Then you are a speedy sight hound like the greyhound, Pharaoh hound, or Saluki. If, on the other hand, you are never in much of a hurry to get anywhere, you would be a bloodhound or a basset hound.

Are you a nervous, excitable type of person who is always tearing about and can never relax for a moment? Are you always sticking your nose into things? Are you an investigative, curious type? If so, you are probably a Parson Russell terrier, Cairn terrier, or some other breed of terrier. If, however, you are the laid-back type who never gets too worked up about anything (except possibly food), you're still a basset hound.

Are you a workaholic who needs to be doing a job to be happy and fulfilled? You rarely take a vacation and wouldn't know what to do with one if you did. You are willing to do any kind of work, no matter how challenging. You love a challenge! And you love the reward that comes from a job well done. If so, you are a Rottweiler or Komondor. Maybe you also love to be in the snow while doing your work. If so, you make a perfect Alaskan Malamute or Siberian Husky. If you like to also take a little nip from the keg now and then, you are a St. Bernard.

Are you a highly organized, take-charge kind of person? Perhaps you're a banker, a teacher, or the CEO of a large corporation. Do you feel the need to constantly direct people, grouping them together for the most effective outcome in your organization? Then you are without a doubt a herding breed like the Australian cattle dog, a collie, or some other breed of sheepdog or shepherd.

Perhaps you're the high-maintenance type of material girl who loves to shop until she drops. You love wearing fancy designer clothes and lots of *bling*. You're undeniably a fashion plate, and money is no object, as long as you are living the good life and have all the comforts you feel you so richly deserve. You could easily be a model, an actress, or a rock star. You're accustomed to being pampered and adored, and people constantly fawn over you and tell you how beautiful you are. And you know they're absolutely right, doggone it! If these are your attributes, then you are a toy breed like a Pomeranian or a Maltese. People had better not toy with your affections, though, because you'll accept nothing but the very best, and that's no Shih Tzu.

Maybe you're just an all-around good sport. You're the outdoorsy type who loves to be out in nature among wildlife. You love a bracing cold morning in the woods, wandering up hill and down dale, taking in the pine-scented air. You might be a hiker, an angler, a skier, or a hunter. If you're happiest while reveling in the great outdoors, you are a Golden retriever, Labrador retriever or other kind of retriever. You could also be a spaniel or a setter. But if you are a pointer, always be careful where you point, especially when you're hunting quail.

Maybe you are just an average, all-purpose sort of person, like the majority of people in the world. You're an all around hard-working stiff, and that is your most admirable quality. You may be the descendant of hardy immigrants with strong roots in the Old World. Perhaps you don't feel you fall into any particular class, but you don't mind that a bit. Being from the Old World, you prefer not to be grouped into a class. You are not beautiful by most people's standards. In fact, you may even look kind of odd and have a name that no one can pronounce. If that is you, then you are a Spinone Italiano or possibly a Löwchen.

Trying to guess what kind of dog you are may have seemed to be kind of a silly exercise, but there is a more serious intent that underlies it. The fact is that too many people choose a dog for all the wrong reasons. Folks still regularly succumb to the "How much is that doggie in the window" mentality and allow their hearts to overrule their heads when choosing a canine companion. Whatever dog you fall in love with will hopefully become your constant companion, so you want to be certain to make the right choice. It's one you'll have to live with for quite some time. The decision to purchase or adopt a dog is not one that should be entered into hastily or unadvisedly—kind of like marriage. In fact, you may be living with that dog a lot longer than you live with your spouse. Considering that studies have recently determined that 31 percent of women spend more time with their dogs than with their husbands, that says a lot about the importance of choosing the right dog, as well as the right husband.

Before you bring home a dog, you'll need to consider your own personality and lifestyle. Some self-examination will help you determine if you are someone who would be better suited to share your home with a lively terrier, an outdoorsy sporting or herding breed, a slow sniffer or speedy sight hound, or a totable toy breed. There are hundreds of breeds to choose from, which can be overwhelming to anyone trying to choose the perfect breed for their lifestyle and temperament. Making the right choice is as important for the dog as it is for you. The reason that so many dogs, including those purebreds of questionable origin obtained from backyard breeders and pet shops, are surrendered to shelters is that someone didn't stop to consider that their adorable little puppy would grow into those big paws or understand what he would be like to live with once he was fully matured. So this is *your* test. It's one every prospective dog owner should be required to pass with flying colors before they bring a dog into their heart and home.

Carefully studying the attributes of breed personality and intelligence will help any potential dog owner make an educated decision. There are many books available at your local library or bookstore that provide detailed information about every breed of dog in existence. One of the best of these is *The Complete Dog Book,* which is the official publication of The American Kennel Club. Their Web site, *www.akc.org,* also offers a wealth of information about the breeds, where to find reputable breeders, the latest dog-related news and events, and anything else you might want to know about your preferred breed. You can also apply your newly acquired knowledge of breeds to choosing from all the equally wonderful, lovable dogs that are available for adoption through pounds, shelters and rescue. Personnel at adoption facilities can help you to make the right choice by providing you with some of the dog's history if it is available. They can advise you whether the breed you are considering is typically energetic or low-key, what kinds of health problems they might have, or if they are good with children. Even if the dog you intend to adopt is a mutt, if you have done your homework beforehand and know your dogs well enough, you can easily determine the predominant breed in the dog's physical and behavioral characteristics, which can also give you a good clue about his IQ.

> "Why, that dog is practically a Phi Beta Kappa. She can sit up and beg, and she can give her paw—I don't say she will, but she can."—Dorothy Parker

Chapter 4

Testing 1, 2, 3

"I bought a dog the other day . . . I named him Stay. It's fun to call him: "Come here, Stay! Come here, Stay!" He went insane. Now he just ignores me and keeps typing."—Steven Wright

A practical test for canines would be to test dogs on their ability to do what they were bred for. But we, as their presumed masters, like to judge things on our own terms, even if it's a different species that is being required to measure up to high-falutin' standards of intelligence.

Beyond Book Smart

Perhaps chasing cars paid off for Duke, a border collie/ Australian shepherd mix, who jumped into a stack of tires over nine feet off the ground and retrieved a toy at the bottom.

The test that is currently in favor among experts asks dogs to perform a series of tasks that might also be expected of a two-year-old child, the estimated intelligence level of most dogs in human terms. As you test your dog, keep in mind that each dog is different. From the affenpinscher to the xoloitzcuintli (the national dog of Mexico, but don't ask me to pronounce it), every breed has evolved for a different purpose. Okay, I admit that's giving the dogs a head start in the game to help them succeed. Isn't that what any parent would do for her kid?

Beyond Book Smart

Nothing better than a clean canine... the largest number of dogs washed in eight hours by a team of twelve people is 848. If that doesn't make you think the next time you complain about washing your dog, what will?

Here are some fun, easy tests that will help you determine your dog's brainpower. These tests have not been confirmed by scientists in a laboratory but by me and my large group of dog-loving friends at home. For a more serious set of tests that will gauge dog smarts broken down into recognizable intelligence classifications, see Stanley Coren's book, *The Intelligence of Dogs*.

There are probably quite a few humans who couldn't pass these tests, so your dog's performance may amaze you and make you one proud dog parent. If he doesn't do so great on the tests, don't sweat it. Remember that he would still love you, no matter how low you scored on the Stanford-Binet. And, of course, no dogs or humans were injured while testing for this book.

Don't let your dog know he's being tested. He might get pre-test jitters, chew up his Number Two pencil, and flunk the multiple-choice. If he thinks it's all a silly game, especially one that's designed from a dog's point of view, he'll have more fun. You will, too. Relax. This isn't the Harvard entrance exam.

Beyond Book Smart

Boxers are widely known for their drooling problem. In Portugal, Gonçalo Veiga's dog, Gorri, is no different. Gonçalo frequently uses a paper napkin to clean his dog's mouth. One day, Gorri was watching Gonçalo's cousin eating in front of the television. As usual, the drool was flowing from Gorri's jowls. Her arm draped over the chair, the girl held a napkin in her hand. She was surprised and amazed when Gorri walked over and wiped his mouth on the napkin. Miss Manners would be proud!

The entire test (six tasks in all) should take no more than an hour to complete, but if you suspect your dog suffers from ADADD (adult dog attention deficit disorder), you might want to let him take the test in small bites instead of administering one long, tiring exam. After all, dogs don't get to enjoy Spring Break.

Beyond Book Smart

Fetch must take on a whole new level when you play with Augie, a golden retriever who holds the world record for most tennis balls held in a dog's mouth at one time--with five balls. Even if he's not book smart, we know he's athletic!

It might be good to have a partner to help you with some of the tests in case your dog needs to be restrained from scarfing down all the treats before you've even begun testing. You don't want your test subject so loaded down with biscuits that he loses his appetite, or his interest. And no fair retesting Fido to get a better score.

Points For Gauging Your Dog's Intelligence

Lassie level—5 points
Eddie level—4 points
Spot level—3 points
Satchel level—2 points
Quincy level—1 point

Test #1—Go Figure

This test measures your dog's ability to figure things on his own.

Set up: You and your dog should preferably be in a space big enough for him to move around without knocking anything over. The only prop you'll need for this test is a towel and a doggie treat.

Goal: The dog should be able to figure out where you've hidden the snack all by himself.

1. Tell your dog to sit (if he can't sit, you don't need to do any further testing, you've got a Quincy).
2. Put the dog treat on the floor and place the towel over it so that it completely covers the treat.
3. Say, "Okay!" and let your dog out of his sit.

Lassie: He moves the towel with his teeth, folds it up, and gently places it to one side. Then he eats his snack. 5 points. He gets an extra 2 points for laundering the towel afterward.

Eddie: He paws at the towel, frantically trying to snag his prize but cannot find it. 4 points.

Spot: He sniffs around the towel, whining and looking at you in confusion. 3 points.

Satchel: He barks and growls at the towel. 2 points.

Quincy: He lies down on the towel and promptly falls asleep. 1 point.

If your dog didn't score in the Lassie category, don't despair. You could always try it again using a really tiny washcloth instead of a towel. I won't snitch to the test proctor.

Test #2—How Quickly Can Your Dog Learn a New Word?

This test measures your dog's ability to recognize new vocabulary words, but your dog won't need a dictionary for this test.

Set up: You and your dog should preferably be outside so that you don't break your favorite vase or crack a mirror, resulting in seven years of bad dog behavior. The only prop you'll need for this test is a ball.

Goal: The dog should learn a new word (no matter how silly it is) and be able to respond to the new word within 10 or so minutes.

1. Decide on a new name for our dog's toy. For instance, you have a ball that your dog knows as "ball." Just for now, decide to call your ball "Fred." Throw your dog's toy or ball a short distance. Your dog should run after it (most dogs will run after a moving object, but if your dog doesn't you might have an easily recognizable Quincy on your hands).
2. Once the dog reaches the toy or ball and touches it, say "Fred."

3. Clap your hands, and show general enthusiasm. Much cheering.
4. Repeat 10 times.
5. Throw the toy or ball and say, "Fred!"

Lassie: She runs after Fred (the ball, silly) and retrieves Fred, bringing it back to you quickly, and gracefully. 5 points.

Eddie: She runs after Fred, grabs Fred and keeps on running. 4 points.

Spot: She runs over to the neighbor's house, who happens to be named Fred. 3 points.

Satchel: She sits at your feet, looking up at you in confusion. 2 points.

Quincy: She's fallen asleep on the towel from the previous exercise. 1 point.

If you don't think that your dog is watching and studying your actions, here's some scientific proof. To test social learning in dogs, a group of Hungarian scientists conducted a series of experiments that specifically studied the effect of a human demonstrator on the performance of dogs. In these tests, they recorded the behavior of dogs when an object such as a favorite toy or food was placed behind a V-shaped fence.

The scientists studied how long it took the dog to obtain the desired object. After six trials, the dogs did not show significant improvement in figuring out how to get around the fence on their own. But after watching a human being demonstrate the skill, the dogs' performance improved within two to three trials.

It made no difference whether the human was the dog's owner or a stranger. The dogs consistently imitated the human behavior they observed. You don't really have to be a scientist to understand that this species we dubbed Canis familiaris, which has coexisted with us two-leggers for the last 100,000 years, learned long ago to play "Doggie see, doggie do" with their human companions.

Test #3 Doggie See, Doggie Do

In this test, we're going to see if your dog can do observational learning and is as smart as Roscoe Jones, who we mentioned earlier (see page 40).

Setup: Purchase a small garbage can with a pedal that you step on to open the lid. The garbage can should be empty when you begin this test. Next, get a doggie treat out of the cookie jar.

Goal: Your dog should learn from your actions and mimic them.

1. Put your dog in a sit.
2. Making sure your dog can see what you are doing, show him the snack in your hand and then step on the pedal so that the lid of the can flies up. When the lid opens, drop the snack into the can. Move your foot and wait for the lid to come down and close tightly over the top of the can.
3. Repeat 3 times.
4. After the third observation, step away from the can and let him have his way with it.

Lassie: He looks at you with a big grin, steps over to the can, pounces on the pedal, opens the lid, puts his nose gently in and retrieves the snack. He winks at you and walks away. 5 points—Score!

Eddie: He runs over to the can, noses up the lid, and enthusiastically chomps on his prize. 4 points. Pretty smart dog you've got there!

Spot: He runs at the can at full speed, knocks it over, and frantically chases the rolling snack around the room. 3 points for enthusiasm and effort.

Satchel: He barks at the can and pees on the floor. 2 points.

Quincy: He's still asleep on the towel you left on the floor from Test 1. 1 point.

If your dog doesn't mimic your behavior around the house, he probably has a lot in common with most husbands. Before retesting, you might have to explain in great detail how basic chores—like throwing laundry in the actual basket—work.

Test #4: Red Light, Green Light

This is an easy test that requires no special preparation.

Setup: Make sure you have a little room so your dog can move forward and you can stay put. All you need for this test is a leash and a bowl of dog food. And your hungry dog, of course.

Goal: To test your dog's ability to obey commands and practice self-control.

1. While your dog is on a leash, make sure the dog sees the full bowl of food. This shouldn't be a problem.
2. Move back about 15 feet from the bowl, keeping the dog on the leash.

3. Before the dog moves to the end of the leash, command the dog to get down. You want your dog to lie down several feet from the bowl and sit quietly until you tell him to get up. Amazing self-control and patience for Fido!

Lassie: Your dog does this off-leash after skipping meals for 2 days because she's been busy rescuing Timmy from the well. 5 points.

Eddie: He gets down, stays down, and stares at you until you deliver human food. 4 points.

Spot: Runs off to find Dick and Jane, dragging you behind him at the end of the leash. You should have used a longer leash. Come, Spot, Come!

Satchel: Checks his wristwatch, Handy, to see if it's really time for dinner. He'd rather eat the cat's food, anyway. 2 points.

Quincy: You guessed it, still asleep on the towel. Poor Quincy can't even stay awake long enough for dinner. 1 point.

Observation

It's 6:00 A.M. and still dark outside. Time for you to leave for work. You're dashing around the house, scarfing down a bagel as you search for your keys. Distracted, you're oblivious to the fact that your actions are under careful surveillance. Out of the darkness, two eyes watch you. They fix upon every step you take as you prepare to leave the house, and they'll be waiting for you to return, ready to pounce the minute you walk in the door.

Who is your mystery mole? Relax, it's not a stalker, the FBI, or the CIA. It's your dog. Your dog always seems to know your every move, even before you make it. It doesn't matter so much whether you understand what your dog is saying to you because he understands everything you say to him, either from your verbal or physical cues.

Although your dog's sense of sight is not his strongest sense (unless he's a greyhound or other breed of sight hound), he's still keenly aware of his surroundings. Any changes in his environment are immediately noticed. Even from inside your house, your dog can spot the squirrel scrabbling across your back fence that you weren't even aware was there. You think he's blissfully snoring away, and then he startles the heck out of you when he suddenly leaps up from his sentry post of sofa or chair and barrels out his dog door, bawling like a hound in a Louisiana bayou. The subtlest movement tips him off that there's an intruder in his territory that must be dealt with. It's that old prehistoric prey instinct at work (although most dogs wouldn't know what to do with that squirrel if they caught it).

Test #5: Buddha Dog

Basically, this test measures your dog's patience and self-control as well as her ability to stay focused and obey commands. It's a fun test to try at the vet, or any other time you want to keep your dog occupied and focused.

Setup: You and your dog should be facing each other. All you need is a pocketful of treats.

Goal: To show your dog that patience has a payoff and to see how patient he can be. There are plenty of humans who probably couldn't pass this test on delaying gratification.

1. Tell your dog to sit in front of you so that you can hold a hand out at eye level with your dog.
2. Take a treat out of your pocket and place it in the palm of one hand.
3. Tell your dog to stay and count to ten.
4. Say, "Okay, Grasshoppa!" and let her have the treat.

Lassie: Closes her eyes, goes into deep samadhi, and doesn't notice when you say, "Okay, Grasshoppa." You fall asleep waiting for her to return to Earth. 5 points.

Eddie: Stares at you with his third eye until you hand over the treat. 4 points.

Spot: Reads your palm while he's waiting for the treat. 3 points.

Satchel: Barks frantically at you, turning in circles and chasing his tail. Forgets about the treat and runs off to pee in the corner. He can't handle the pressure. Fuzzy the cat eats the treat, instead.

Quincy: He's fallen asleep. Again. 1 point.

"If a dog will not come to you after he has looked you in the face, you should go home and examine your conscience."—Woodrow Wilson

Test #6 Where'd It Go?

This test is designed to gauge how quickly your dog responds to a situation that defies his expectations.

Setup: This test should take place outside where you can throw a ball, a stick, or a toy for your dog. Try to have a fair distance for the dog to run to fetch.

Goal: See how long it takes for your dog to discover the flying object isn't going where she expects it to go. All you need is a toy and a good throwing arm.

1. Take your dog outside and put her in a sit.
2. Throw the ball, stick, or toy and tell her to fetch.
3. When she brings it back for you (or after you retrieve it, Silly Dog Owner), repeat once.
4. The third time, throw the object in a completely different direction.

Lassie: Lassie watches you intently, immediately notices the changes in direction, fetches, returns, and then saves Timmy from the well. 5 points.

Eddie: Eddie watches you throw the ball repeatedly, catches on that you're trying to fake him out, and decides he'd rather chase his tail instead. When he's tired of doing that, he bounds off in the right direction to fetch the ball, but it's probably because he's so dizzy from spinning.

Spot: Spot runs off in the wrong direction, just like an Eddie, but a Spot runs around madly looking for the toy, then barks at you in frustration. You have to point out the object and run and fetch it yourself. He'd rather run after Dick and Jane. 3 points.

Satchel: He finally fetches the ball but then lets Fuzzy take it away from him.

Quincy: Doesn't move a muscle, but waits for you to fetch the toys and give him a snack. Could he be a Lassie in disguise? 1 point.

> "If you think dogs can't count, try putting three dog biscuits in your pocket and then giving Fido only two of them."—Phil Pastoret

Test #7: The Real Test

Ask most any dog trainer and they'll tell you that dogs can learn any dog trick in the book if trained properly and long enough. Therefore the real test is not *What's Your Dog's IQ?* but what's the owner's IQ? Are you smart enough to train a dog? Or did you give up long ago, assuming your dog is untrainable? Do you think Poochy is good enough as is, even if counter-surfing is an Olympic event in your kitchen?

Setup: A bit of hard and fast honesty, an accurate self-appraisal and some humility.

Goal: To accurately assess your qualification to teach an animal how to do anything at all.

1. Sit in a quiet place and think of that behavior your dog just can't seem to "master." Be it stay, down, roll over, stop jumping on the guests, chasing the mailman—whatever.
2. How long did you try to train your dog for this behavior? ten minutes? One day? Two days? A week? Remember, we all make mistakes or do stupid things. I'm sure you can think of a few. Did you give up too soon?
3. Don't assume that your dog is stupid. Most likely you just gave up too soon. The dog might be confused by mixed signals you are giving him. When we are properly motivated (house training, for instance), our dogs seem to learn complex new behaviors within two weeks. We can learn some patience from our dogs. Maybe we should try Buddha Dog on ourselves with dinner!

This time, you get rated:

Lassie: Congratulations! You realized the error of your ways and are planning to call a dog whisperer to train you instead of the dog. 5 points.

Quincy: When it comes to training dogs, you are either a Lassie or a Quincy. Which one are you? 1 point.

Once you have completed the tests, add your dog's scores together. Add Rover's scores from Test #1 through

Test #6. Then check your score from Test #7. Who has the higher IQ?

How to Rate Rover

- **30 points:** Canine Einstein
- **18–29 points:** Pretty doggoned smart
- **7–17 points:** Not the brightest pup in the pack
- **Below 6 points:** It shouldn't happen to a dog

How to Rate You

- **You got 5 points:** Congratulations! Give yourself a treat. Oh, you already have one: a wonderfully well-behaved dog.
- **You got 1 point:** Change your legal name to Quincy and get your dog a sweater that says, "I'm with Stupid."

By now you should have completed all the tests, rated how well your dog performed, and determined where he ranks in intelligence. If your dog scored well on the tests, congratulations! However, if you're still waiting for your dog to wake up from on top of the towel, you have my condolences. Perhaps you can send him to summer school.

Chapter 5

Boost Bowser's Brain Power

"It's funny how dogs and cats know the inside of folks better than other folks do."—Eleanor H. Potter

So your dog didn't do so well on the tests I suggested, did he? Well, don't hold it against him. It may not mean he's stupid but just understimulated. If your dog is anything like mine, he's more accustomed to lying around like a dog-skin rug and being fed on demand than having to earn his rewards. He may have been perplexed and even a little cheesed off about having to actually perform a task to obtain a treat. There are lots of ways to improve his performance on this and other intelligence tests without sending him off to remedial dog school, although in some cases that's probably not a bad idea.

This chapter presents many options for exciting doggie adventures that may be the change he needs.

Training Classes

The most common method of improving your dog's mental mettle is through obedience training. If you're the competitive type, you can enter trials and compete with other dogs. According to *Dog Owner's Guide, The Online Magazine for All Pet and Showdog Owners* (*www.canismajor.com*), there are many levels at which your dog can compete:

Novice obedience: The dog is judged on his ability to heel (on and off leash), stand, stay, and come to the handler when called.

Open obedience: The exercises done within this level are done off-lead and include the following:

- Heeling pattern
- Figure eight
- Down on recall
- Retrieving on the flat and over a high jump
- Broad jump
- Group exercises performed with the handler outside of the ring

Utility: This class is far more difficult than the previous two levels of obedience training and is not for the casual canine competitor. Training dogs to compete at this level is rigorous and demands just as much training for the handler as the dog. Of the dogs that do compete, the brainier breeds excel in utility, but that doesn't mean your dog can't advance to this level of competition with hard work and proper training. In this class, the dog is required to perform a complex heeling pattern off lead, which includes hand signals that direct him to obey commands to go down, sit, come, and return to heel. The dog must also seek and retrieve various articles with his owner's scent (no, not socks and underwear) from a pile on the floor of the ring as well as do high jumps and bar jumps.

Rally obedience: Rally obedience, or Rally-O, is a fun new sport that is best compared to rally sports car racing and dog agility, where dogs speed through a timed course. As with the other classes, dogs can earn titles, according to their ability. Obviously, some speed breeds will be better at this than others.

Non-regular classes: This is the Everydog class, which is often included at trials and intended mainly for fun and practice. It includes the following levels:

- Prenovice level, where all exercises are done on leash.
- Graduate novice level that combines novice, open, and utility exercises.
- Braces, or pairs of dogs of the same breed working novice routines together.
- Veteran level for senior dogs.
- Versatility level for dogs trained through utility level.

Unfortunately, some dogs never receive any obedience training. Some aren't even allowed to venture outside their own back yards. If you never challenge your dog's intellect or expose him to new experiences, how can you expect him to not to be a dullard or a dunce? Dogs are like children; they need regular attention and stimulation. A solitary life is a boring life, and it's no life at all for a pack animal like the dog. Even Robinson Crusoe had Friday to keep him company when he was stranded on that remote tropical island. Your dog needs company, too. Preferably yours. Participating with your dog in training classes and competitions provides not only valuable behavioral instruction and structure but also much-needed social interaction with other dogs and people. You also might make some new friends among other dog lovers. Because dogs are by nature social animals, they teach us to be more social, too.

Beyond Book Smart

In 2004, Striker, a border collie, broke the world's record for the fastest time a dog has unwound a non-electric car window. Smart and quick—does it get any better?!?

All the training you give your dog can't alter his basic personality, any more than your parents' efforts at training changed yours. In any litter of pups, you'll likely observe that one is naturally more bold or timid than the others. Hold the bold pup on his back, and he will struggle and fight to right himself, while the timid pup will be more docile and offer no resistance to being forced to lie on his back. As to which puppy in the litter will make the better companion, it depends again on which dog will match your lifestyle and personality. In most cases, the pup that is neither too bold nor too timid usually makes the best pet.

Out and About

Dog parks, dog-friendly beaches, and other off-leash facilities are also excellent outlets for socialization and recreation for your dog. Dogs even go away to camp nowadays. Summer camp probably wasn't half as exciting for kids when you were growing up as it is today for dogs. The activities are designed to engage and challenge your dog. There are even canine arts-and-crafts classes. Best of all, the dogs don't get homesick while they're away at camp and beg their Muddah and Faddah to let them come home.

Bark in the Park

If you hear a cacophony of barks in the park, it's likely you are in one of the many off-leash dog parks that are popping up everywhere across the country. As leash laws have become more stringent in urban areas, the need has grown for spaces where dogs can be exercised without worrying about traffic or being fined by the pooch police. Many cities, including my own, have been slower than an old hound dog to designate adequate leash-free space for dogs to exercise. For example, in California's capital city, there are only three dog parks, and none is larger than two acres. That's a pretty small area in which to contain the numbers of dogs and owners that frequent the parks.

Beyond Book Smart

Labrador retrievers, golden retrievers, German shepherds, collies, rottweilers, and scores of mutts played a large role in the search-and-rescue operations at the World Trade Center after the tragic events of September 11, 2001. Bravery is not limited to humans, as these heroic dogs proved.

The whole dogfight over off-leash space began when county parks officials started ticketing owners who let their dogs run off-leash along the American River Parkway, as they had done for decades. Sacramento Dog Owners Group (SacDOG) was formed in 2002 in a grassroots campaign for more dog recreation areas, and progress is being made. After years spent in a tug of war between SacDOG and the Sacramento County Parks Department, a whopping seventy-five acres of open space was earmarked in 2005 for an off-leash recreation area. Dog owners are howling for joy! Other cities should follow suit in order to provide their citizens' pets with a safe and clean place to play.

Of course, enjoying off-leash areas requires some responsibility from dog owners, too, and park rules must be obeyed to make the park fun and safe for everyone. First and foremost, owners must scoop the poop. When you get large numbers of dogs in a confined space, this task becomes paramount. If you forget to bring a bag (we all do sometimes), most dog parks provide bag dispensers (as well as other amenities like water fountains and double gates, so your dog can't escape the area).

Your dog should be well trained and obey your commands. Leashed or not, nobody likes being around a dog that is badly behaved, so don't unleash your problem dog on others at the dog park and give the rest of us a bad name. Your dog must be socialized with other dogs. It's surprising how many people fail to accurately assess their dog's sociability with other dogs or people. Just because a dog is friendly to you doesn't mean he's friendly with everyone. There have been some unfortunate incidents in parks where dogs were attacked and seriously injured by someone else's dog. If there are children present in a park, this is of vital importance, since some dogs do not react well to the noise and activity of children.

Dogs should be fully immunized against diseases before being exposed to other dogs. This is especially important for puppies, who may not yet be protected against diseases such as distemper or parvovirus.

Beyond Book Smart

In Hollywood, California, a Canine Heroes Walk of Fame is popular tourist attraction. Perhaps a Nobel Prize for pups is next!

Always pay attention to your dog. Dog parks are not just great social gathering places for dogs but also for people. It's easy to get distracted in conversation and lose track of your dog's whereabouts. Keeping an eye on your canine kid can avert problems before they start.

Life Is a Dog Beach

Ah, the seaside! There's nothing like getting sand between your dewclaws at the beach, cavorting in the breaking waves, and barking at the seals that are barking back at you. It's an experience every dog should be able to enjoy at least once. Unfortunately, there are restrictions about taking dogs on many public beaches, mainly because people don't like getting other things besides sand between their toes. At the least, you'll be required to keep your dog on a leash, which isn't nearly as much fun for you or your dog.

It's getting harder to find a dog-friendly beach, especially near large metropolitan areas. In fact, according to recent research, dogs have been banned entirely on the majority of beaches throughout the United States and Canada. So, before you set out for a day at the beach with your pooch, you'd better check to be sure that the beach allows dogs. If it does, will your dog be allowed to run leash-free, or are leash laws enforced? Fines for violators can run $250 and higher, so do your research before you storm the beach with Bowser.

Beyond Book Smart

Bear, a golden retriever, was one of the first "officers" to respond when the twin towers came under terrorist attack in New York City.

Truthfully, dog owners themselves are largely at fault for the diminishing leash-free beaches and other dog-friendly areas. Stiffer leash laws have been imposed because dogs have been allowed by irresponsible owners to run amok and bother other people and their dogs and soil the sand where children play or sunbathers lie. I have lost count of how many times I've stepped in something exceedingly unpleasant at a public park. The few offal offenders spoil it for others. It's every dog owner's responsibility to ensure by their and their dog's behavior that the few remaining dog-friendly beaches continue to stay that way.

The Web site *www.dogfriendly.com* recommends the following beach etiquette for dog owners:

- Always keep your dog leashed when there is a leash law.
- Always clean up after your dog by using a poop bag or pooper-scooper.
- Do not let your dog visit with other beach-goers or dogs, unless welcomed.
- If using an official off-leash area, your dog needs to be well behaved and must listen to your verbal commands.
- In an off-leash area, always pay attention to your dog.

Bone Appetit

Continental canines have been accepted in and about eateries in Europe and Great Britain for years. Seeing dogs inside English public houses is as common as dartboards and draught beer. You might even see a dog sipping a pint when his master is distracted by the barmaid or throwing darts. French poodles sampling *pomme frites* (the treat Americans once dubbed Freedom Fries) in sidewalk cafés on the Champs Elysées is trés chic in Paris, but Americans have been slower to set an extra place at the table for American dogs.

"To his dog, every man is Napoleon; hence the constant popularity of dogs."—Aldous Huxley

Because of stricter health regulations in the United States, most restaurants still do not permit pets inside or even in outdoor seating areas (service and guide dogs excepted). Since our canine friends have a tendency to be *au naturel* in their elimination habits, "Pass the Grey Poupon" could take on a whole different meaning in such a venue, which is why the rules are rather stringent about dogs around dining establishments. But happily, this doggie discrimination is slowly disappearing. More and more often, dogs are being permitted to accompany their owners' *al fresco* dining experiences. One can only say, Bone Appetit!

"No animal should ever jump up on the dining room furniture unless absolutely certain that he can hold his own in the conversation."—Fran Lebowitz

Fido Fast Food

With so many people taking to the road with their dogs, fast-food restaurants could soon be adding something new to the drive-up menu—fast food for Fido. Could McDoggies be far off? How about Jack in the Barks? Most dog lovers would probably agree the idea is not too far fetched. In fact, one former fast-food executive, Kim Buchanan, came up with a treat called Fetch Fries for dogs.

You probably wouldn't be a bit surprised to learn how many people place an extra order at the drive-thru for the fur kid begging in the back seat. You've probably done it yourself. It may not be too long before you see menu selections like Big Mutt Mac or McFurry at the Golden Arches.

In the meantime, there are countless barkeries, pawtisseries, and doggie dining establishments across the country that are devoted specifically to canines. Three Dog Bakery, a franchise with stores all over the United States and in Japan and Korea, also specializes in tasty treats for dogs. Browse online through their "Dogalog" (at *www.threedog.com*) to order everything from biscuits and bits to gift boxes and celebration cakes. You can even order monthly Dogliveries of yummies for your dog's tummy.

Home, Sweet Home

Sometimes it's kinder to leave your dog behind when you travel, especially if the dog is old, doesn't travel well, or the trip will be too stressful, not just for the dog but for you, too. If you will be in a situation that is emotionally charged, such as a funeral, it could be distressing for dogs, who are by their nature highly sensitive to our moods.

There are still some places that are not dog friendly, and owners should think about how a pet will be received by others. Let's face it: Not everyone is a dog lover. Pity those poor unenlightened creatures. But if fish and houseguests stink after three days, as the saying goes, then a houseguest with a dog in tow stinks a lot faster, especially if that dog is ill mannered or poorly housetrained. If your dog is liable to chase the cat, bark at the parrot (which may bark back), nip at the kids, anoint the $3,000 designer sofa, or mark the rare Oriental rug (even the best trained dog slips up every now and then in unfamiliar territory), or if your host is allergic to dogs, make other arrangements for your fur friend while you visit. It'll be more considerate for everyone, and you might actually be invited back. The following sections explain some of the alternatives to Fido tagging along.

Hello Muddah, Hello Fido

Pitching a pup tent takes on a whole new meaning when you take your dog to one of the many dog camps that have sprung up across the country. You can find many dog camps on the Internet or in pet magazines like *Fido Friendly,* which monthly features dog friendly places across the United States and abroad.

Beyond Book Smart

When **doghero.com** announced a call for entries for Top Five True Stories of Dog Heroics in 2004, they received over two hundred submissions!

Camps such as Camp Winnaribbun, located on Lake Tahoe, offer campers the opportunity to participate in activities like obedience, agility, herding, tracking, and flyball as well as games, crafts, and photo sessions. It also provides health services such homeopathy, psychocybernetics, massage therapy, and, of course, first aid. Those of us who ever went off to camp as kids know it isn't without its occasional scrapes or insect bites. Like the summer camps of our youth, this one has the rustic cabins to sleep in after a day filled with nature hikes, arts and crafts, S'mores, and, finally, storytelling around a roaring campfire. Best of all, it's a peaceful getaway for your dog and his best friend.

Just as with any camp, there are rules to abide by. For instance, Camp Bow-Wow, a dog-camp franchise with many locations nationwide, requires that its canine campers be at least four months old, spayed or neutered (if over six months), and current on vaccinations for rabies, distemper, and bordatella (canine cough). Dogs must be in good health, flea and tick free, well socialized with other dogs, and playful.

Doggie Day Care

Have you ever seen anything sadder than the look on your dog's face when you have to leave for work? (Your face is probably almost as sad, come to think of it.) If you're like most people, you have to work to bring home the bacon-flavored treats. Unless you have a home-operated business or you telecommute, you must leave your dog behind while you work. When left to their own devices for eight hours at a stretch, dogs can do some serious damage to the house, particularly if they suffer from separation anxiety. Sometimes even having two dogs to keep each other company in your absence is no help. They just become partners in crime.

This is why doggie day care has come to be in such great demand. When you consider that many people can barely afford day care for their human children, it gives you a clue about how elevated the dog has become in our culture. We must keep Fido happy and well adjusted at all costs!

Beyond Book Smart

Assigned to JFK International Airport, Crazy Joe, a yellow Labrador rescued from a shelter, has found more than ten million dollars worth of cocaine, heroin, and other narcotics. His handler rewards Crazy Joe's big busts with a juicy steak. Looks like Crazy Joe isn't so crazy after all!

Fortunately, most doggie day-care centers are not as expensive as their human counterparts. Owners may pay as little as $5 an hour for their pets to participate in such activities as games, pool splash, story time, arts and crafts (paw painting), and field trips to local parks.

Puzzles

While your dog obviously can't work the New York Times crossword (who can?) or assemble a 1,000-piece jigsaw puzzle, there are other kinds of puzzles designed for dogs that can help boost his brainpower. For starters, repeating the tests described in Chapter 4 once in a while can improve his performance. Practice makes perfect.

There are many new interactive toys on the pet market that challenge your dog's problem-solving abilities. Treat balls are a favorite of owners and dogs alike. One of the most popular of these is the Kong treat dispenser.

A toy that takes this concept a step further is the Talk to Me Cosmic Treatball, which the manufacturer describes as "light years ahead of its time." The Treatball can not only record your voice in a message for your pet to comfort him while you're away, but when the dog rolls the toy, it dispenses treats and displays a dazzling light show that engages all the animal's senses. The lights are vision safe, so they won't dazzle your dog senseless.

Play with Me!

There are also games and activities you can devise to play with your dog. Spending quality time with your fur child is the best thing you can do for him, and it might just make him smarter—at the least, it will seem that way. Playing games together improves communication with your dog and develops mutual respect and understanding. They can be fun, too. Here are a few fun and simple games recommended at the Web site *www.dog-play.com*:

Follow the Leader: Scatter traffic cones or other obstacles randomly about and have your dog follow you through the maze. This is even more fun when done with other dogs and owners and you take turns being the leader. Some dogs may need to be led through the maze, but eventually they may get the idea they're supposed to follow you. Here's where those obedience classes can pay off!

Find the Treat: This is similar to an Easter egg hunt, but you get to skip the reeking rotten eggs that manage to remain hidden under the sofa or in your bureau drawer for months after. When your dog's not looking, hide some of his favorite treats around the house or yard, and then tell him to go find them.

Hide and Seek: Tell your dog to sit and wait while you go hide. Make sure he doesn't peek. Then call your dog and see how long it takes him to find you. If you haven't hidden up in the attic, he should be able to sniff you out fairly quickly.

Agility

Agility is just one of the many sports dogs and their owners are participating in. This is also a maze of a sort, in which dogs navigate their way over, under, and through various obstacles including hurdles, seesaws, and tunnels. Over time, dogs improve their speed and time running an agility course. There are also many agility-based games, like Snooker, that are enjoyable and challenging for both dog and handler. Snooker is loosely based on the popular English tabletop game of snooker, only it's played with colored flags instead of colored balls. In snooker, the dog maneuvers through sequences of jumps and obstacles to earn the most points. The game is about having the best strategy to beat your opponent and how to collect the most points in the shortest amount of time.

Agility is exciting and challenging for the dog, and it's not so bad for their owners, either, since they have to move as fast and efficiently as their dogs do through the often-complex courses. It requires practice and repetition to compete in agility trials. Agility will improve not only your dog's mental and physical fitness but also your own.

Beyond Book Smart

Jenner, a Golden Retriever, is the loyal, skilled guide dog of John Hart, a visually-impaired resident of San Francisco. A dog about town, Jenner assists John with his community service and volunteers with him at the VA Hospital to put patients at ease.

Traveling Dogs

When Willie Nelson sings about being on the road, he's singing about traveling with the boys in the band, but more people are taking to the road and the skies with their dogs. Globetrotting canines are more the rule than the exception these days, and airlines, hotels, and motels have necessarily become much more dog friendly than they used to be.

Airborne Dogs

When you're taking a long trip, sometimes the only way to bring Fido along with you is to fly him to your destination. According to the U.S. Department of Transportation, over 2 million pets and live animals are transported by air each year. More and more dogs are flying the dog-friendly skies and becoming world travelers as they accompany their owners to destinations near and far. Airlines are getting on board with the flying dog trend and going the extra mile to make the trip safer and more comfortable for your best friend. In fact, Jet Blue Airlines was the winner of the 2005 Five Dog Bone Award from *Animal Fair* magazine for its pet-friendly accommodations. Smaller dogs can ride right along with you in the cabin, and the airlines take greater care in the handling of pets than was once the case. New federal laws have been passed that make airlines more responsible for the care and safety of your animal. However, many people still prefer to leave their dogs in the care of a reliable sitter than subject them to the noise, temperature changes, and stress of traveling in the cargo hold of a jet plane.

United Airlines has even established a special Pet Class and offers frequent flyer miles and other perks for your pet. Midwest Airlines treats every frequent-flyer Fido at check-in with two yummy organic carob chip cookies and a spiffy crate tag. Your dog can also earn a free round-trip ticket for every three round trips, six single flights, or by redeeming 15,000 of your frequent-flier miles.

Obviously, not every dog can travel in the cabin with you sipping Cocker-tails and watching *Best in Show* on the in-flight movies. You definitely can't fit a sky kennel for a Great Dane under the seat in front of you, and that means larger dogs will have to travel in the cargo hold. Because of the dangers of transporting pets in cargo holds, the Humane Society of the United States recommends against it unless it is absolutely necessary to do so. Their organization has received countless complaints of animals that have been lost or injured or that have died from extreme heat or cold, suffocation due to lack of oxygen, or rough handling while in airline cargo holds. It is largely due to such complaints that airline companies have adopted more stringent regulations, and cargo transport for pets has been greatly improved.

Highway Hounds

A less stressful alternative to flying for your pet is for him to ride along with you in the car. Car travel is also much safer for your dog than it once was, thanks to comfortable crates and more dog-friendly cars with canine restraint systems, which protect your dog from injury in case of sudden braking or an accident. The place not to carry your pet is on your lap. It not only hampers your ability to drive safely, but if your airbag should deploy it could seriously injure your pet. He'll be much safer cruising with you in a crate or a harness.

Room Service for Rover

Hotels more frequently accommodate pets nowadays. Some even roll out the red carpet for Rover. And if he pees on it, not to worry. Some of the more posh hotels even provide canine room service with T-bone steaks on the menu, massage, personal dog walkers, and even a biscuit on the dog's pillow.

Beyond Book Smart

Shug, a six-year-old golden retriever, is an Animal-Assisted Activities/Therapy Dog. Affectionate, patient and caring, he is an indispensable resource in the hospitals where he helps make patients feel less fearful and lonely during their stay. Heroes comes in all shapes and sizes—and breeds.

When you travel with your dog, finding dog-friendly accommodations is always a consideration, but no one wants to stay at a fleabag hotel. Fortunately, that's no longer a problem. Loews, Omni, Red Roof Inn, and Motel 6 are just a few of the hotel chains that have joined the Welcome Waggin'. Even five-star hotels and resorts worldwide will roll out the rollaway bed for your dog, although that wasn't always the case.

"A couple of years ago, there were very few hotel chains that would allow pets," says Shawn Underwood, a communications representative for PETCO. "Now they realize they're missing out on the market share because so many people are traveling with their pets. If they don't allow pets, those people are going to go somewhere else."

Countless Web sites, including *www.letsgopets.com,* are now devoted to finding pet-friendly accommodation and other helpful pet-travel information. Most hotels charge a deposit for doggie guests to cover any carpet cleaning, flea bombing, or repairs to damage after you leave, and some have size restrictions. If your room is as small as some I've stayed in here and abroad, sharing it with a large-breed dog could be like having a bullmastiff in a china closet.

People travel in order to gain new experiences and expand their horizons, and they want no less for their pets. Seeing the great big world out there is bound to be a lot more stimulating for your dog than staring at the walls of a kennel, and he won't experience separation anxiety if he's traveling along with you.

Chapter 6

Dumb and Dumber Dogs

"Dogs come when they're called; cats take a message and get back to you later."—Mary Bly

It's easy to experience some disappointment if your dog didn't do so well on the tests I suggested, has never helped a wheelchair-bound person across the street, or has never rushed to rescue Timmy from a burning barn. If Spot just stood there with the towel over his head and didn't budge an inch the whole time you tested him, or if his idea of agility training is not falling in the toilet while he drinks out of it, don't fret. Maybe he just didn't understand that you wanted him to escape the confines of the towel and then run off to find the cure for cancer. After all, there wasn't a squirrel racing across the fence or a cat running up the tree to make him want to fling that dumb blanket off his head and give chase. The stopwatch hasn't been made that's fast enough to clock the average dog pursuing a squirrel or a cat that happens to wander into his back yard. Now those are the kinds of stimuli dogs can really get excited about.

Even if your dog failed the test, that doesn't mean he's still not the best doggoned dog since Rinty bit Black Bart in the breeches. Smart or dumb, we love them just the same. Even that aviating novelist, Snoopy, had to return to Daisy Hill Puppy Farm now and again for a refresher course.

As any dog owner will attest, sometimes it's hard to tell who's running the show.

Dog Rules

1. The dog is not allowed in the house!
2. Okay, the dog is allowed in the house, but only in certain rooms.
3. The dog is allowed in all rooms, but he has to stay off the furniture.
4. The dog can get on the *old* furniture only.
5. Fine, the dog is allowed on all the furniture, but he's not allowed to sleep with the humans on the bed.
6. The dog is allowed on the bed, but only by invitation.
7. The dog can sleep on the bed whenever he wants, but *not* under the covers.
8. The dog can sleep under the covers by invitation only.
9. The dog can sleep under the covers every night.
10. Humans must ask permission to sleep under the covers with the dog.

"My Name Is 'No, No, Bad Dog!' What's Yours?"

When it comes to training a dog, you need to accentuate the positive, eliminate the negative. Rewarding good behaviors is kinder and more effective than punishing bad behaviors, as the title of this section, borrowed from Gary Larsen's cartoon, illustrates so well.

"If I have any beliefs about immortality, it is that certain dogs I have known will go to heaven, and very, very few persons."—James Thurber

When you command Rover to roll over, he does it because he desperately wants that tasty treat you are holding up before him as an incentive to perform that task. He completes the task, and you praise him saying, "Good boy, Rover!" and then hand over the treat. It's likely that next time you command him to sit or roll over, he'll do it because he knows that he's likely to receive praise or a reward for his effort.

Beyond Book Smart

The difficult task of searching the affected areas of the Pentagon after it came under terrorist attack was assigned to, among others, fifteen certified rescue dogs.

We like being rewarded or getting a pat on the back for work well done, too, don't we? And we're likely to want to do a good job next time if we know we'll be praised for our efforts. Maybe Rover's a lot more human than we give him credit for. And there's no doubt he's a lot smarter than we might think. Some experts claim that dogs don't really love us, at least not in the way that we define love. But they love the food and treats we give them. Perhaps they've just been playing us all along, and we were too stupid to notice. Even a dumb dog is likely to obey a command if the promise of food is in the bargain. If you engage your below-average dog with games and toys that involve food, you may be surprised at how intelligent he suddenly seems to be.

> Don't accept your dog's admiration as conclusive evidence that you are wonderful.—Ann Landers

Train Humane

Choke, pinch, spike, shock. No, I'm not talking about going ten rounds in the ring with Mike Tyson but about widely used methods of dog training. Are these the kinds of things you should do to your best friend? The San Francisco Society for the Prevention of Cruelty to Animals (SFSPCA) doesn't think so. According to the SFSPCA, choke and shock collars may stop the dog from pulling or barking, but that's only because *it hurts*. In some cases, they say the dogs not only do not respond to the collars but also become desensitized to them to the point where injurious levels of force are required to make the dog respond at all. More often than not these pain chains are used incorrectly, making them ineffective and cruel. In the SFSPCA's opinion, strangling a dog to the point of damaging its trachea is not training; it's abuse. If humans were subjected to some of the same "training" methods that are used on dogs, it would be called torture.

Beyond Book Smart

In April of 2003, Buster, a springer spaniel, saved thousands of civilian and military lives when he sniffed out a cache of arms, drugs, and explosives in Southern Iraq. Talk about smart use of a nose!

The SFSPCA believes that it is possible and preferable to train a dog without inflicting pain and that halter and reward-based training is far more effective and beneficial to the animal. That's what they use to train their shelter dogs, and they say that the dogs make progress much more quickly than with other methods. A well-trained and humanely trained dog is an adoptable dog.

During break time at obedience school, two dogs were talking. One said to the other, "The thing I hate about obedience school is you learn **all** this stuff you'll never use in the real world."

Training Is a Snap with a Click

Dogs are masters at reading our body language. They constantly pick up on cues, both visual and auditory. Dogs can even mimic human body movements and expressions such as shaking hands, slapping a high-five (or is it a high-four?), smiling, singing, and even dancing, one of the newer dog freestyle sports.

Your dog knows when you're leaving the house almost before you do. He immediately notices your hurried pace as you dash about the house collecting your things. He hears the telltale jingle of your car keys, and when the suitcase comes out of the closet, it's time to bar the door! He gets that hangdog expression that makes you feel so guilty because he knows that this time you're leaving him for more than just a few hours.

Beyond Book Smart

Buck, a mountain cur like Old Yeller, defied the odds and survived being trapped in a 70-foot sinkhole for sixteen days in the Great Smoky Mountains Park. He barked and barked until Ranger Rick Brown found him and rescued him. It just goes to show: you can't keep a good dog down.

The dog is a master empathizer. He knows when you're happy, sad, or angry and even seems to share those emotions with you. Have you ever had a dog place his paw on your hand or respond to you in some other way when you're distressed? If you had a dog during the bumps and bruises of childhood, you have known the comfort a canine companion can give. Some dogs have been known to laugh or to cry real tears right along with their owners.

> "Dogs understand your moods and your thoughts, and if you are thinking unpleasant things about your dog, he will pick it up and be downhearted."—Barbara Woodhouse

We all know how keen those ears are. From the other end of the house, your dog not only hears and responds with gusto to the front doorbell (or even one on television) or the clank of the mail slot, he also hears the squeak of the refrigerator door opening or the faintest crinkle of plastic wrap, no matter how quietly we try to unwrap that block of cheese. What better candidate could there be for clicker obedience training?

The clicker is one of the most effective methods of training used to reinforce a positive behavior. According to trainers at K-9 Insight Obedience, of Planet Pooch in Redwood City, California, clicker training is based on the principle of operant conditioning, which forms an association in the dog's mind between a behavior and its consequences. When the desired behavior is performed, the trainer presses a hand-held clicker that emits a high-pitched sound, which serves as a marker. When the dog hears the clicker snap, he knows that he did something good and that a reward is at hand (or rather in his master's hand).

> "Trained or not, he'll always be his own dog to a degree."—
> Carol Lea Benjamin

If there is ever any doubt that your dog is smarter than you give him credit for, consider this dictionary of terms—and the *real* definitions of the words, as proven by our dogs every day.

Doggie Dictionary

Bath: This is a process by which the humans drench the floor, walls, and themselves. You can help by shaking vigorously and frequently.

Bicycles: Two-wheeled exercise machines, invented for you to control body fat. To get maximum aerobic benefit, you must hide behind a bush and dash out once you see a human riding one. Then bark loudly. Run alongside for a few yards until the person swerves and falls into the bushes, and you prance away.

Bump: The best way to get your human's attention when he or she is drinking a big full cup of coffee or tea.

Deafness: This is a malady that affects you when your person wants you to come inside and you want to stay out, or vice versa. Symptoms include staring blankly at the person, then running in the opposite direction or lying down.

Dog bed: Any soft, clean surface, such as the white bedspread in the guest room or the newly upholstered couch in the living room. However, avoid the round, pillowy looking thing your owner calls the "dog bed." This is no good whatsoever.

Drool: Is what you do when your people have food and you don't. To do this properly, you must sit as close as you can while they're eating. Look sad, and let your saliva fall to the floor, or better yet, on their laps.

Garbage can: A container that your neighbors put out once a week to test your ingenuity. You must stand on your hind legs and try to push the lid off with your nose. If you do it right you are rewarded with margarine wrappers to shred, beef bones to consume, and moldy crusts of bread.

Goose bump: A maneuver to use as a last resort when the regular bump doesn't get the attention you require. Especially effective when combined with the sniff. See *bump* on previous page.

Lean: Every good dog's response to the command "Sit," especially if your person is dressed for an evening out. Incredibly effective before black-tie events.

Leash: A strap that attaches to your collar, enabling you to lead your person where you want him/her to go.

Love: A feeling of intense affection, given freely and without restriction. The best way you can show your love is to wag your tail. If you're lucky, a human will love you in return.

Sofa: We use sofas for the same reason that humans use napkins. After eating, it is polite to run up and down the front of the sofa and wipe your whiskers clean.

Sniff: A social custom to use when you greet other dogs. Place your nose as close as you can to the other dog's rear end and inhale deeply, repeat several times, or until your person makes you stop.

Thunder: This is a signal that the world is coming to an end. Humans remain calm during thunderstorms, so it is necessary to warn them of the danger by trembling uncontrollably, panting, rolling your eyes, and following at their heels.

Wastebasket: This is a dog toy filled with paper, envelopes, and old candy wrappers. When you get bored, turn over the basket and strew the papers all over the house until your person comes home.

Stupid Is As Stupid Does

There are some common reasons that your dog may not be as dumb as you think he is, and they are reasons that are often overlooked by frustrated owners. Your dog could be bored, sick, or just plain stubborn.

Room and Bored

Your children receive report cards in school that grade them on how well they perform and whether that performance meets the expected standards for their grade level. Children usually try to meet the expectations of their teachers and parents, and the same is true of our dogs. If we expect little or nothing of them, that's pretty much what we get. Dogs also need to be challenged and engaged in interesting activities to be their best. If your dog sits around the house alone all day long, the chances are that he's not dumb—he's bored.

Dogs also need the companionship of their owners. There's not much point in having a dog if you are never around to give him any attention. A burglar alarm doesn't eat its weight in dog chow or shed on the furniture. The more quality time you spend with your fur kid, the better he'll do on tests like the ones in this book.

Sick As a Dog

If your dog doesn't do much of anything, it's possible that he could be ill or have some other physical problem you haven't yet diagnosed. Here are some of the more common signs of illness, according to the University of California, Davis School of Veterinary Medicine:

Fever: When was the last time you took your dog's temperature? Do you even know how? For most people, using a rectal thermometer on a dog rates right down there with expressing his anal glands, but taking your dog's temperature is a simple step you can do at home before consulting the veterinarian. Normal body temperature ranges from 100°F to 102.5°F.

Pain: This is the most obvious sign something is wrong. But since dogs can't talk, people don't always know when a dog is in pain. Obviously, if he yelps for no reason or is limping, he's probably in pain. Dogs in pain will withdraw from normal activity and hide or lie down in unusual places when experiencing pain.

Lethargy: An observant owner will notice immediately if his dog is less energetic than usual or is resting or sleeping too much. If you have a basset hound or certain other low-energy breeds, lethargy is not necessarily a sign of illness.

Vomiting: Probably the least fun part of owning a dog, vomiting can just be due to a mild tummy upset. If it's persistent, this can signal a serious medical problem, especially if accompanied by some of the other symptoms listed above.

Changes in appetite and body weight: If your dog's appetite or body weight increases or decreases, it could be a sign of an illness such as diabetes. Be warned, though, that if he's putting on weight consistently over time, it's probably because you're overfeeding him. If you use the free feeding method, he might also be eating out of boredom.

Most people are quick to notice changes in their dog's appearance or behavior. If you're not sure about the symptoms you observe, always consult a veterinarian.

Stubborn As a Mule

If you think mules are stubborn, you should see some breeds of dogs that are inherently more stubborn, which accounts for why some are less trainable than others. Stubborn dogs don't do as well on tests like those in Chapter 4 and, as a result, their intelligence often gets a bad rap. People are quick to label stubborn dogs as stupid dogs, but it's more accurate to say that they have a mind of their own. Many breeds of hound, terrier, and other hunting dogs have a stubborn streak, which is really more determination than stubbornness.

> Living happily with stubborn dogs all depends on your outlook. You can choose to be annoyed by their lack of compliance to your commands, or you can just accept this independent behavior as another endearing quality of the breed.

If you've ever tried leading a basset hound in a direction he doesn't want to go, you quickly learn that it's much less stressful for you and more pleasant for the dog to simply go with the flow. If stubborn dogs could talk, they'd tell you that's not a bad approach to life in general.

Smarter Than They Appear: Hidden Talents Your Dog May Have

While people have always been pretty sure they know what makes a dog smart (Does he come when he's called? Go lie down when he is told to do so?), dogs appear to have other abilities that suggest types of intelligence never before expected in the species.

The Color Pupple

Can dogs see colors? Well, no one has ever asked a dog whether or not he sees the world in living color, but recent evidence has altered the long-held opinion that they see only in black and white.

> Our basset hound, Butterscotch, was particularly fond of a blue racquetball we called Bluey. As time went on, we added another racquetball to Butter's toy collection, this one in green. She soon learned to tell the difference between Bluey and Greeny and would fetch whichever one we asked her to. This convinced me that she was able to tell the difference between the two colors.

Dogs have dichromatic vision, which means that they can see only a portion of the range of colors in the visual spectrum of light wavelengths. It is a common belief that dogs lack the ability to see the range of colors from green to red but can see in shades of yellow and blue, a sort of canine colorblindness. Humans, on the other hand, have trichromatic vision, which means that they can see the entire color spectrum.

Dogs have an advantage over us, though. They can see in much dimmer light than we can, which is why they never trip over their dog bowls in the dark. The center of the dog's retina is composed of rod cells, which perceive shades of gray. By contrast, the retinas of humans are composed primarily of cone cells, which enable us to perceive color. Rod cells require much less light to function than do cone cells.

Beyond Book Smart

Bob Hope's USO tour had nothing on Boris, a Belgian Malinois that has been a military canine for the U.S. Armed Services for nine years. He has served in Bosnia, Kosovo, and Iraq, where he boosts morale for homesick soldiers.

Dogs can also detect motion much better than we can. When out in the field, a dog can spot the slightest movement in the brush that might signal the presence of game. That's why dogs notice that sassy squirrel on the fence long before you do. By the same token, your dog is able to hone in on the tiniest morsel of food that might drop from your dinner plate to the floor.

Although your dog may be watching every move you make and everything else that moves in his doggie universe, he doesn't see things with the same sharp, clearly defined edges that we do. He can't focus as well on the shape of objects. What we can see clearly may appear blurred to a dog that is looking at the object from the same distance. Compared to human beings, dogs have a visual acuity of 20/75, meaning what they can see at twenty feet a human with normal vision can see clearly at seventy-five feet. Unless someone comes out with corrective lenses or doggie contacts, your dog's vision will never be quite as good as yours. Still, there's no disputing that his ability to see into the heart of man is always a perfect 20/20.

Can Dogs Tell Time?

Kathy Hoxsie swears her Norwegian elkhound, Daisy, can tell time. She says that if Daisy's second meal of the day isn't served to her promptly at 3:30 P.M., she begins to talk. Likewise, Daisy visits Kathy every evening precisely at 8:00 P.M. to say good night.

Another testament to a dog's ability to tell time (admittedly sillier and made up) is the cartoon strip Get Fuzzy. In this cartoon, the comical shar-pei named Satchel has his own wristwatch, which he calls Handy.

The first person to study the dog's ability to measure time was the famous physiologist Ivan Pavlov. In his famous experiments on conditioning, Pavlov trained dogs to wait to salivate before receiving food. In addition to proving that dogs can be conditioned to slobber on cue, he proved that dogs are probably better than most humans at delaying oral gratification (probably why there are no Jenny Craig weight-loss programs for dogs).

Actually, dogs don't tell time in the strictest sense, unless you happen to have one of the new robotic dogs like Aibo or Poo-chi, on which you can actually set an alarm.

The measurement of time is a uniquely human concept, but according to biological studies done at the University of Utah, animals do behave according to defined schedules. In the wild, these schedules are associated with the likelihood of obtaining food. If your dog knows he'll be greeted with a biscuit every time you come home from work, he soon learns how to predict with amazing accuracy when you'll return to the house each day, so in a sense he can tell time. And he doesn't even need to have a wristwatch handy.

Common Reminders for the Less-Than-Intelligent Dog

Regardless of how your dog fared on the tests in this book, or what your plans are to make sure he scores better next time, if you have a less-than-intelligent pup on your hands, you may want to teach him to commit these rules to memory—for both your sakes.

- I will not play tug-of-war with Dad's underwear when he's on the toilet.
- The garbage collector is *not* stealing our stuff.
- I do not need to suddenly stand straight up when I'm under the coffee table.
- I will scootch my bottom along the grass (rather than the carpet) to rid myself of hangers-on.
- I will not roll my toys behind the fridge.

- I will shake the rainwater out of my fur *before* entering the house.
- I will not drop soggy tennis balls in the underwear of any person sitting on the toilet.
- I will not roll my head around in other animals' poop.
- I will not eat the cat's food, before or after it has been digested.
- I will stop trying to find the few remaining pieces of clean carpet in the house when I am about to throw up.
- I will not throw up in the car.
- I will not roll on dead birds, seagulls, fish, crabs, or other smelly things.
- I will not lick my human's face after eating animal poop.
- "Kitty-box crunchies" are not food.
- I will not eat any more socks and then redeposit them in the back yard after processing.
- The diaper pail is not a cookie jar.
- I will not wake Mommy up by sticking my cold wet nose up against her back.
- I will not chew my humans' toothbrushes and not tell them.
- I will not chew crayons or pens, especially not the red, or my people will think I am hemorrhaging.
- When in the car, I will not insist on having the window rolled down when it's raining outside.

- We do not have a doorbell. I will not bark each time I hear one on television.
- I will not steal my Mom's underwear and dance all over the back yard with it.
- The sofa is not a face towel. Neither are Mom and Dad's laps.
- My head does not belong in the refrigerator, dishwasher, or trash can.
- I will not bite the officer's hand when he reaches in for Mom's driver's license and car registration.
- I will not spend more than five minutes trying to find the perfect place to poop.
- I will not take off while on leash to chase squirrels while Mommy is standing on a slippery grass slope.

Chapter 7

Tails from the Bark Side

"If dogs could talk, it would take a lot of the fun out of owning one."—Andy Rooney

Long before Mr. Ed, the talking horse, whinnied his first words to Wilbur, people have wanted to understand animal language. This is especially true of dog lovers. Over the past 10,000 years or so, humans have learned to read canine body language with some degree of success—or have we? Considering that there are almost 5 million dog-bite victims in the United States every year—not always from someone else's dog—this may not be the case. While we certainly can become accustomed to our own dog's behavior, it is never a safe assumption that we know exactly what is going on in the canine mind.

A smarter concept to understand, rather than spend your time trying to figure out your dog's every thought, is that we don't always know what our dog, or the neighbor's dog, is thinking. We assume a wagging tail means the dog is friendly, but are we acting presumptuously? Experts tell us that whether the dog is friend or foe depends more on the position and speed of the tail wag, the set of the ears, and other body signals. Of course, the dog's primary means of communication, and the one we should pay most attention to, is its bark.

Beyond the Bark

When your dog barks, you'd better listen. He may be trying to tell you something. Just as we human beings speak in many different tones of voice depending on the circumstances, so do dogs. Those differences are immediately recognizable to those who live around dogs daily and become accustomed to the subtle variations in their language. When I'm working in my office, I can tell by the tone of my dogs' barks whether I should keep on typing or if there's a situation that requires my prompt attention. (Of course, in the minds of my dogs, I'm sure everything is a Code Red on the dog alert scale.)

Dogspeak

Are you fluent in Dogspeak? What is your dog really saying when he barks? If he lives in Massachusetts or

Mississippi, does he speak with an accent? Dogspeak is a difficult language to learn, and as far as I know, there are no classes in how to speak dog.

Beyond Book Smart

Arthos, a stalwart Beauceron search-and-rescue dog from Germany, helped save the life of a suicidal twelve-year-old girl by leading police to the distressed child.

Some researchers at the University of California, Davis, recorded the barks of ten dogs of varying breeds and found that various pitches of canine vocalization seemed to have different significance. They discovered, for instance, that a single high-pitched bark meant, "Where has my owner gone?" while a lower-pitched, harsher bark suggested, "A stranger is approaching."

If you listen closely to your dog's vocalizations and observe the situations in which they occur, you soon will be fluent in the language of Dogspeak and be able to recognize the various barks that convey messages like these:

- "I'm hungry. Feed me now."
- "Call of nature. Let me out before I piddle on the floor." (For the universally understood "yark-yark" sound, no translation is required. Better run on the double before your dog yarks his kibble up on your carpet.)

- "There's a squirrel in the tree. I wouldn't know what to with it if I caught it, but I'd like to chase it, anyway."
- "The neighbor's cat is on the fence. I'd chase it, but I'd probably get clawed to ribbons.
- "Animail!"
- "There's an ax murderer on the front porch dressed as the meter reader."

No matter what your dog's intelligence level, he always knows what he means when he barks, even if you don't. In fact, they understand our language much better than we understand theirs. If, like most people, you're not very good at deciphering dog language, you could always buy one of those clever Bow-Lingual gadgets to help you translate what Rover is saying.

A Bark Is Worth 1,000 Words

A couple of Christmases ago, my brother gave me a Bow-Lingual Dog Translator—the ultimate gift for a dog nut like his sister, he was no doubt thinking. I was thrilled! At last I would finally know what my dog was saying to me when he barked. The battery-operated Bow-Lingual is easy enough to use. First you enter some information about your dog, such as size, sex, and breed. Then you attach a wireless microphone to your dog's collar. When the dog barks, a handheld device displays the translation. The device even has a "home alone" mode so you can monitor your dog's barks and behavior while you are away. Other modes offer training tips and a health check to help alert you to signs of illness in your pet. Some Bow-Lingual translations of your dog's vocalizations are "Resistance is futile" and "Are you my friend or my enemy?" When Bubba barked at me, the translation was, "Give me the sun and the moon." No surprises there.

Wag the Dog

Besides paying attention to the sound of your dog's bark, pay attention also to his body language. These are clues to what he's trying to tell you and the emotion he is expressing with his vocalizations. Is his stance stiff or relaxed? What is the position of his ears, erect or flat? With lop-eared dogs like bassets and bloodhounds, it's admittedly hard to tell the set of the ears, but there are noticeable changes if you're paying close attention.

Is the dog's tail wagging or not? Is it carried high or tucked? A dog's body language is just as important a signal to humans as the bark (or growl) in knowing whether it's safe to approach or to back off. It would behoove everyone to take time to recognize those signals and to learn to decipher the language of Dogspeak.

Stance

If a dog's stance is stiff, he's asserting his dominance. This is one of the first physical indicators that a dog could become aggressive toward another dog or a human. When you take your dog to a park and he encounters another dog, you've probably noticed that the dogs approach each other and assume this dominant body language. The neck is arched and the tail is held high. The tail tip may be wagging slightly. Staring is also a dominant behavior that may accompany the stiff-legged stance, and it has nothing to do with who gets the last bagel.

Beyond Book Smart

Harvey, a three-year-old bulldog in London, England, wasn't arrested for speeding, reckless driving, or car theft when he bumped his owner's $60,000 Maserati into gear and floored it into a parked van. The insurance company refused to pay because Harvey wasn't listed as a driver on the owner's policy. "I thought the light was green. Whaddaya expect—I'm colorblind!"

After a little butt-sniff boogie, the dogs might decide they are equal in status and that it is time to be friendly with each other. Alternately, the behavior may escalate. The dogs might begin exhibiting more dominant behavior, such as growling and baring of teeth, and you may have to haul out the garden hose when a fight suddenly breaks out. One dog may even try to mount the other dog, which also shows dominance, doggie-style.

As in the wild, domestic dogs usually quickly resolve such potentially injurious disputes when one dog backs down, adopting submissive body language such as rolling on the back and exposing the belly or licking at the other dog's face, which is Dogspeak for "I give. You rule!" You can observe this pack hierarchy behavior even in a litter of puppies. Deciding on the identity of the pack leader is a very important aspect of being a dog.

Ears

Other than his nose, the ears are a dog's most important sensory tool. They are also a good indicator of his mood. When a dog is on high alert, his ears are erect. Dogs with pricked ears use them like radar to detect the faintest of sounds. The tips of the ears (or pinnae) pivot to better determine the direction from which sounds are coming. In the wild, dogs can detect danger or prey long before it can be seen. That's why you probably sometimes wonder what the heck your dog is barking at when you can't see a doggoned thing. It's because he's hearing sounds that are far above and beyond your auditory range. When a dog is relaxed, his ears are also relaxed. Of course, with long-eared dogs like bassets the ears are always relaxed, just like the rest of the dog. The fact that the ears hang down over the ear canal also someone inhibits the dog's ability to hear as keenly as short-eared dogs. The sound is necessarily muffled. The dog's hearing is even further impaired if he is wearing a snood to keep his ears from dragging in his food bowl.

Tail Tales

The tail language of a wild dog is different from that of a domesticated dog. The wagging tail is generally considered a sign of friendliness, but as we've seen, that is not always the case. According to *The Body Language and Emotion of Dogs,* by Myrna M. Milani, D.V.M., tail-wagging behavior varies greatly from breed to breed. In this instance, "the tail wagging the dog" is probably pretty accurate. Dogs have been conditioned by humans to wag their tails according to the work they do. For instance, certain hunting breeds are expected to wag their tails constantly while tracking their quarry and then hold their tails rigid to signal the hunter that they've spotted the prey. Sled dogs carry their tails high, while herding breeds hold theirs low. A lowered or tucked tail can also signal submission, especially if the dog rolls on its back and piddles. Some dogs don't even have tails! So you have to be something of a tail reader to know exactly the long and short of what a dog might be trying to say to you.

"Dogs laugh, but they laugh with their tails."—Max Eastman

What do you do if you encounter a growling dog whose tail also happens to be wagging? Which end do you believe—is he friend or foe? The tail may be held high with just the tip wagging. The tail may even be wagging more vigorously, which further confuses things. The dog might also exhibit that stiff stance and some of the other telltale signs of a potentially unfriendly dog. When the dog bares his teeth at you (no, he's probably not smiling), the tail goes rigid and the hair on the dog's shoulders spikes like a punk rock hairdo, things could get ruff!

Walk, don't run. Running excites the dog's prey pursuit instinct, and you may soon be feeling a draft when the seat of your pants is missing. Back away slowly. Don't make direct eye contact with the dog, which could be construed as a threat. This scenario is typical behavior in the pack in which a dog asserts its dominance over another dog, but it also extends to humans. If more people were aware of the warning signs of aggression in dogs, there would be far fewer mauling incidents in the United States. As for me, if it's a choice between heads or tails, I'll always bet on heads.

> A dog can express more with his tail in minutes than his owner can express with his tongue in hours.

Can Dogs Really Talk?

Dr. Doolittle and Ace Ventura talked to the animals, and many other people these days say that they do as well. While there is no way to know for sure whether these people are truly communicating with canines, there are many people out there who insist they are serving a valid and real purpose by speaking with dogs. Cindy Huff, who practices the special arts of Animal Communication, Massage, and Healing Touch is one of the people who talks to animals, primarily dogs. Huff has been communicating with animals professionally for two years, but she discovered her ability to communicate with animals as a child. She talks mostly to dogs, predominantly basset hounds and greyhounds, through rescue organizations such as Guardian Angel Basset Rescue and Greyhound Pets of America. Her own bassets, Beau and Maggie Mae, are rescue dogs. Lately, Cindy has been crossing sides, as she says she's talked to dogs that have recently passed over.

Huff says, "I look at myself as a bridge builder and teacher, someone who bridges a language gap between animals and their humans, someone who also teaches humans basic communication techniques with the animals they love." Huff sees the goal of her business as helping animals and their human companions better understand each other. By creating a conduit between humans and their pets, animal communicators help relay to owners any troublesome issues from the animal's perspective. Owners also can learn how best to address those problems. "It is our hope that by promoting this understanding, the relationship between humans and their companion animals will become more fulfilling, more joyful, and more satisfying for all involved." In the case of rescued dogs, where the animal's history is unknown and there are often health or behavior issues, being able to communicate with one's pet becomes especially valuable.

Kindred Spirits

"I've grown up with animals my entire life and was very close to all of them," says animal communicator Mary Argo, who works mostly with domestic animals such as dogs. "I knew when things were wrong with the animals, but I assumed everyone could pick up on these feelings." She says it all came together for her when she read J. Allen Boone's book *Kinship with All Life*. "That book changed my life," Argo says. "I decided from then on that communicating with animals was what I wanted to do."

Beyond Book Smart

Drew Barrymore rescued Flossie, a yellow lab mix, years ago. Flossie returned the favor when she awakened the actress and her then husband, Tom Green, in time to escape a devastating house fire by barking and banging on their door. Ms. Barrymore rewarded Flossie by placing their two-story Beverly Hills home, valued at $3 million, in trust for her heroic dog. That's some doghouse!

Argo says she deals with a wide variety of issues or problems people might have with their dogs. "Many consultations are about how the animal is feeling physically," Argo says. "I make it very clear that I am *not* a vet and I cannot diagnose, but I can help the person understand how the animal is feeling and perhaps refer them to a vet who might be able to help."

"No one appreciates the very special genius of your conversation as the dog does."—Christopher Morley

Both Huff and Argo say they ordinarily conduct their work over the telephone, using a photo or the owner's detailed physical description of the animal and then making a "heart connection" with it. Some communicators even work via e-mail, so again there is no body language or any physical reference to work from, but the results are still amazingly accurate, astounding even skeptics. So how does it work?

Huff explains, "I receive emotions, pictures, physical sensations in my body (especially when asking them how they are feeling in their body), sounds, tastes, words, fleeting nuances. It depends very much on the animal."

Argo and Huff also sometimes conduct sessions with the animal physically present. At these sessions, which usually last between thirty and sixty minutes, the communicator typically picks up body language cues as well as vocalizations to augment the telepathic signals they receive from the animals. "With most dogs, I can interpret some vocalizations—but not all," Argo says. "Some dogs love to hear themselves talk, and they vocalize just to hear the sound of their own voice."

Both women said they prefer to be called animal communicators, not animal psychics, a term many people find intimidating or off-putting because it implies that animal communicators are soothsayers who make predictions or have supernatural powers other people don't have.

"I try to impress on my clients that anyone can talk to their animals and that it just takes practice and not special skills," says Argo. "It's very similar to learning a new language." Becoming *bow*lingual, you might say.

Whisper in My Ear

If you have ever seen the movie *The Horse Whisperer,* starring Robert Redford, you probably already have a good grasp of what a dog whisperer does. A whisperer is called in for a crisis situation and is often the owner's only hope of keeping the traumatized animal. As with horse whisperers, dog whisperers can help you to see the world through your dog's eyes and understand why he behaves the way he does. More important, you'll learn how to correct and control unwanted behaviors in your pet, which is different from the type of work Cindy Huff and Mary Argo do. Animal communicators such as Huff and Argo communicate with dogs, but not for the specific purpose of training them, as do dog whisperers.

> "If you take a dog which is starving and feed him and make him prosperous, that dog will not bite you. This is the primary difference between a dog and a man."—Mark Twain

Known for his National Geographic television series *The Dog Whisperer*, Cesar Millan has been called the Dr. Phil for dogs. Millan says he helps the dogs he works with to retrain their owners rather than break the dog of a particular behavior.

Remember, there are no problem dogs, only problem people. Millan employs a "power of the pack" approach in his training.

Millan and many other people who work with dogs tend to encounter the same problem in dog owners. People lavish love and attention on their dogs without first expecting them to do anything to earn it. That puts the dog in control. By requiring the dog to perform a task before receiving his reward, the owner establishes dominance as the leader of the pack. The dog will strive to please his person and be happier and better adjusted. That approach has often been known to work with kids, too.

Here are some training tips based on those shown on *The Dog Whisperer*. Note that it's recommended to consult a professional in matters of training problem dogs (or people):

1. The entire family should be involved in the decision to bring a new dog home. The way in which responsibilities for the dog's care are to be shared should be established before the dog arrives.
2. The breed of dog you choose should fit your lifestyle.
3. Maintaining the health and happiness of a dog is a responsibility. Before you commit, ask yourself if you're ready for that responsibility. Do you have the time or patience to do what is required for a dog?
4. Some dogs come with a history. (Remember my Crazy Daisy?) When adopting an older dog, be aware that he may have had negative experiences in his former home that could affect his reactions toward people, kids, or other animals.
5. When you bring home a new dog, make him a priority. He expects and deserves your time and attention. Take time out daily to establish rules, set boundaries, exercise him, and to give him affection.
6. Give the dog something to accomplish before you share food, water, toys, or affection. Rewards should be earned.

7. Make time for exercising your dog daily for at least forty-five minutes.
8. To show your dog who's boss, always be the first one to walk in or out of a door. Don't let him barge ahead of you.
9. It can be a drag being pulled down the street at the end of the leash, especially if you're a human. Your dog should follow the leader—*you*! Don't let your dog lead you when walking him. Keep him beside you or behind you. That way he knows who's leading whom. Someone please tell my dog!
10. Keeping a dog costs money—sometimes a lot of money if a medical emergency arises. Put aside some cash in your doggie bank and budget for your dog's care, training, and other expenses. Consider purchasing pet insurance to help with veterinary bills. Most vets demand payment at the time services are rendered.

The number of U.S. households with a dog is at 43.5 million and rising. Learning to better understand your dog can only serve to enhance your relationship with him. Surely, if we expect our dogs to understand us, we should attempt to better understand our dogs. Whether you do that through an animal communicator, dog whisperer, or your own instinct, take time to get in touch with your inner dog. Your best friend will thank you in ways no human ever could.

Chapter 8

Do Dogs Express Emotion?

"The Dog was created especially for children. He is the god of frolic."—Henry Ward Beecher

Because we have become so closely bonded with our dogs, we humans have a tendency to believe that our canine companions experience the same emotions we do. Why should we not believe it? Dogs certainly seem to feel joy, pain, fear, and grief, and those are all human emotions. Whether we actually share similar feelings as these creatures that share our lives or whether dogs have just become expert mimics of our behavior to elicit a reward has not been proven. The Oxford Companion to Animal Behavior advises animal behaviorists to remember, "One is well advised to study the behaviour, rather than attempting to get at any underlying emotion." However, any dog lover will profess with certainty that animal emotion is as real as any human's.

Do Dogs Smile?

Do dogs smile? This is an interesting question. If we look at dogs in movies such as *Because of Winn-Dixie,* the answer is yes, although it's digitally enhanced. One time in a dog shelter, I saw a dog grin at me ear to ear just like Winn-Dixie does in the film. I knew she was trying her best to persuade me to adopt her. I have always been sorry that I didn't. I sure hope someone else did, because a smiling dog deserves a good home.

In *The Hidden Life of Dogs* (1993), Elizabeth Marshall Thomas stated that dogs do smile. She said their faces become relaxed and pleasant, just as a human face does when smiling. The dog's ears are low set, his eyes are half shut, the lips are parted and the chin is held high. Some dogs imitate human smiles, baring their choppers like someone told them to say "Cheese" before snapping a photo. Your dog may or may not have grinned ear to ear at you when you performed the test to measure his social learning (in Chapter 4), but such a behavior is certainly a social response in which the dog mimics human behavior and seems to understand exactly what a human smile means. When a human smiles, he's happy. So is a dog. You don't have to be a behavioral scientist to understand that.

A dog has lips, so theoretically a dog can smile. (Okay, some people claim they're just panting, but we dog lovers know better). You may have seen the issue of *The Bark* magazine that featured photos of people's dogs smiling, laughing, and grinning like a Cheshire cat (though I guess that would be grinning like a Cheshire dog, in this case). Dogs even seem to have a sense of humor. It's nice to know that at least someone will laugh at your jokes.

Beyond Book Smart

A German shepherd, Torr, foiled a bank robbery and dangerous hostage situation at the HSBC bank on Buckingham Palace Road in London. Torr pinned one gunman down on the ground and quickly sniffed out the other fugitive, who was hiding in a ventilation shaft. Thanks to this Sherlock Bones, he gave himself up and no one was hurt.

Here are some handy tips for capturing your dog's smile for posterity.

Getting Your Dog to Smile for the Camera

- Choose an appealing background for the photograph, one without the shredded newspaper, chewed shoes, or demolished garden.
- Get down to his level for the shot, but you don't have to roll on your back.

- Use a treat or toy to get the dog's attention. Having a helper do this frees your hands for the camera, and you don't get saliva on the lens.
- Close-ups are best, but don't aim so close you just get the nose and nothing else. You don't want a fish-eye effect, unless you're photographing a fish.
- It's better to shoot the photograph at a three-quarters angle than straight on. It's more artsy-barksy for pup portraiture.
- Take the shot in bright light, and avoid letting shadows fall on the dog. That means you may have to bring him out from wherever he's hiding in hopes of avoiding this photo session.
- Try to capture the pet's individual personality in a pose that is unique to him. Well, within reason, of course. If his usual pose is that of a Mississippi leg hound, better choose another.

Do Dogs Cry?

In *Goodbye, My Lady,* a tearjerker of a dog movie featuring Walter Brennan and Brandon de Wilde, the star was a basenji, the African barkless dog. This strange little dog could laugh and cry real tears, according to the boy, Jesse, who found her lost in the swamp. Whether the dog's tears were the result of real emotion or just a physiological response to swamp gas is hard to say, but I have observed tears rolling down my dog's cheek, usually in response to my own tears. Whenever I cried as a child, my dog rested his paw on mine in what I could only assume was sympathy. Whatever had hurt me was quickly forgotten because I knew that at least my dog understood how I felt. This leads me to believe that dogs must at least feel empathy for our emotions, even if they do not feel the same emotions themselves.

Tears serve the purpose of clearing the eye of irritants and keeping it moist. All land animals are capable of producing tears, but scientists assert that humans are the only animals that cry tears of emotion. Pups and other young animals cry for their mothers. While these cries are not accompanied by tears, there seems to be an underlying emotion.

Beyond Book Smart

When Allen Parton was struck by a car and thrown from his wheelchair, a fast-acting rescuer—Endal, a yellow lab—moved Mr. Parton to safety, covered him with a blanket and brought a mobile phone to him. Endal, Allen's constant companion, was awarded the PDSA Gold Medal for Animal Gallantry and Devotion to Duty.

It's hard to imagine upon hearing the lonesome howl of the wolf that it is nothing more than a mating call and that it does not also contain some measure of woe and despair.

Perhaps it's the same for the canine species as it is with the human male: Real dogs *do* cry.

Do Dogs Grieve?

You may have heard the story of Greyfriar's Bobby, John Gray's faithful Skye terrier in Edinburgh, Scotland, who kept stubborn vigil over his beloved master's grave for fourteen years. Often in bad weather, attempts were made to keep him indoors, but his dismal howls succeeded in making it known that this interference was not agreeable to him, and he was always allowed to have his way. Day or night, he could be seen in or about the churchyard, and nothing could induce him to forsake the hallowed spot where his master lay buried. When the little dog died in 1872, a statue and fountain were erected in his honor.

Anyone who has ever lost a pet knows how much we grieve over them, but our pets grieve, too. Animals grieve for the loss of another animal in the family or for the loss of an owner, either through displacement or the owner's death, as in the case of Greyfriar's Bobby. They also experience many of the same emotions associated with loss that we do. Cats grieve for dogs and the reverse is also true, even if they sometimes fought like cats and dogs. Mother dogs have also been known to grieve after their puppies are taken away from them. They sometimes wander through the house searching for their young. Sometimes lactating females will become surrogate mothers to squirrels, cats, and even pet rats! Some will adopt a stuffed toy that reminds them of the baby they lost. Grieving pets will often seek out the departed companion in the places where they slept or played together. It's not hard to understand these behaviors once you pause to ponder the sense of loss they must feel, which is no different than what we would feel under the same circumstances.

According to Elizabeth Kübler-Ross, the first of a few Western psychologists and sociologists to study the psychology of grief in humans, the stages of grief are as follows:

1. Shock and disbelief
2. Anger, alienation, and distancing
3. Denial and guilt
4. Depression
5. Resolution (closure)

Grieving pets experience symptoms similar to those experienced by a grieving pet owner. They may show signs of stress or restlessness. They may stop eating and lose weight or become listless and depressed. They may have trouble sleeping or even howl pitifully over the loss of their companion, as Bobby did.

Just as it would for you, it will take time for your dog to adjust to the loss and for him to establish his new status in the pack hierarchy. Be patient with the process. Don't introduce another pet too soon after the loss of the other pet. Some claim that your dog might even adopt traits of the deceased pet, but if he starts sniffing catnip or using the litter box, it may be time to get another cat. It's also important to your dog's healing process to maintain his routine in feeding and exercise. Try not to overdo it with treats or attention, or bad behavior patterns could emerge and leave you with a fat, finicky pet.

Beyond Book Smart

Guardian angels have two wings and sometimes they also have four paws. Buoy, a yellow lab, was dressed as an angel for his master's Halloween party. When outside taking a potty break, he discovered Dragica Vlaco, suffering from hypothermia after a fall into the freezing Columbia River. Buoy stayed with her until police found her—just in time to save her life.

You might want to consult an animal communicator, if you are so inclined. They may be able to help your dog (and you) heal from the loss of an animal or human companion.

In her book *Natural Healing for Dogs & Cats* Diane Stein suggests some homeopathic remedies for aiding the grieving process, such as the following:

- Ignatia, a homeopathic remedy that helps a grieving animal (or human)
- Bach Flower Remedy's Honeysuckle with Star of Bethlehem, for the animal who has lost a person

- Essential Essences' Comfort Essence, to help a dog or cat adjust to a new family
- Apricot or Koenig Van Daenmark, for animals whose person has cancer
- Royal Highness helps an animal (or human) prepare and go through the impending loss
- Bleeding Heart, to help the animal or person let go
- Borage, to help with overcoming grief
- Perelandra's White Lightnin', to help with the shock and trauma of early grief stages

Pawltergeists—Messages from Pets Beyond the Bridge

"They're heeeeere!" That may be just a line from a spooky movie about ghostly manifestations, but anyone who has ever lost a beloved pet knows that our pets remain with us, at least in memory. Some animals go a bit further than that. Sometimes they come back!

Anyone who has ever looked into the loving eyes of a companion animal understands that he has an essence and a soul. We understand that the spirit of an animal doesn't die when the light leaves those eyes. Even though a pet no longer has a physical form, his unique soul always remains a part of our lives. Many people believe our pets also send us signs to reassure us and let us know that even though they couldn't stay with us as long as we wanted them to, they are ever present. And just as when they were alive, they have attention-getting ways of making their presence known.

Now, I'm not talking about clanking choke chains or levitating dog bowls, but you may have thought you heard the faint but unmistakable jingle of dog tags that echoes through the suddenly too-empty house. People frequently claim they hear a familiar click of toenails on the kitchen floor or the sound of paws padding across their bedroom in the night. Or you might fancy that you hear the happy "Welcome home" drumming of Fido's tail or feel Fluffy circling your ankles rhythmically to the whirr of the can opener. There is other evidence of the presence of these benevolent pawltergeists, if you're paying close attention.

One theory about the appearance of these furry phantoms is that when owners feel great love and grief for an animal, it causes the pet's spirit to return for a time to console the owner. Since they know how to do this so well in life, it makes perfect sense that they should seek to do the same for us in the afterlife. As we begin to heal from the pain of the loss, the visitations grow less frequent. When we are finally able to let go, they cease altogether.

In a letter to the editor of a local newspaper, a woman wrote in reference to the attention given to pets that were lost in the devastation of Hurricane Katrina that the loss of a pet is not as great as the loss of a human. She wrote that it is shameful to consider the life of a dog or cat as important as that of a person. To her, many pet owners would reply that if that animal is the only companion that you have in your life, as it is for many single and elderly people, the importance of the life of that dog or cat becomes very great, indeed.

"If there are no dogs in Heaven, then when I die I want to go where they went."—Will Rogers

The stages of grief are the same for the loss of a dog or cat, or even a gerbil, as the letter writer mentioned to drive home her point that the life of a pet is of little consequence. I can attest from firsthand experience that the grief over the loss of an animal is as great, and often greater, than the loss of a person.

Beyond Book Smart

Some of the oldest dogs ever recorded include Bluey, an Australian cattle dog who died in 1939 at the age of twenty-nine. Another dog, an Australian cattle dog/Labrador mix, died in 1984 reportedly at the age of thirty-two. Bramble, a collie, celebrated her twenty-seventh birthday in 2002 and ate a diet of rice, lentils, and organic vegetables every evening. Maybe those vegetarians are onto something.

Perhaps it is the knowledge that our time together is finite that makes it all the more precious. One thing is certain: Neither our pets nor we live forever, but the love we feel for them and that they feel for us never dies.

Chapter 9

Career Canines: When Dogs Are More Than Pets

"Histories are more full of examples of the fidelity of dogs than of friends."—Alexander Pope

While every owner thinks her dog is smart in one way or another—whether smart enough to fetch the paper or smart enough to know which member of the family is most likely to give him a treat—there are dogs in society that are undeniably intelligent and that serve humankind with those smarts. One of the benefits about having dogs instead of kids is that you never have to send your dog off to college. But if your dog scored high on the tests in Chapter 4, you might want to reconsider. Maybe Spot would be well suited to attend an advanced obedience class. Or perhaps your dog could be a member of the workforce. There is higher education that goes above and beyond basic training for dogs that display special aptitudes. There are even dogs who are looking out for our well-being every day.

> Kathie Hoxsie says her Norwegian elkhounds are uncanny. Her first elkhound, Lady Dog, could forecast earthquakes that occurred more than 200 miles away. She'd get nervous, pacing back and forth at their Lake Tahoe home several days before an earthquake hit the Bay Area or Mammoth Lakes region. That's a good dog to have around in case of a natural disaster.

Of course, all dogs possess special qualities simply because they are our best friends, closest companions, and protectors. The following sections describe a few of the many beneficial ways our dogs serve mankind.

K-9 to Five—Dogs at Work

While it's hard to envision your dog saving lives as you watch him scratch himself in front of the television, perhaps you have a potential career canine who has some unique talent that could benefit society. The best and brightest of dogs are out among us every day performing invaluable services in search and rescue, assistance, and security. We've all seen the K-9 dogs riding shotgun in the backs of police cars as well as the bomb-sniffing dogs in airports. These dogs lay their lives on the line to save ours. There are many jobs that canines are capable of doing, and the highest profile of them serve our community every day.

A dog walks into the unemployment office and asks a man behind the desk if he would help him find work. The man, astonished at the sight of a speaking dog, replies, "I think I can help you." The guy immediately gets on the phone to the circus to find out if they can use the dog in their show. The dog overhears some of this conversation and says, "I hate to interrupt, but what would the circus want with a brick layer?"

Sheriff Without a Gun

Some public servants have four legs instead of two. You'll see these dedicated officers doing ride-alongs in K-9 law enforcement patrol vehicles. These deputy dogs don't pack heat, but they do wear a badge and a bulletproof vest. Specially trained police dogs perform a broad range of duties, just as their human counterparts do, and the danger to the canine officer is every bit as real.

Some of the services these dogs perform include tracking, apprehending criminals, performing area searches, detecting narcotics and explosives, and assisting in SWAT teamwork. The dogs reduce the risk to human officers by using all their senses, but mostly they use their strongest—the sense of smell.

Beyond Book Smart

Norche, an artistic Scottish terrier belonging to Bettina Rister, is thinking of changing her last name to Van Dogh. The twelve-year-old terrier's Paw-tistic Impressions water-colors were featured on Animal Planet and have sold for up to eighty dollars each. The proceeds benefit the Scottish Terrier Rescue of North Alabama.

Keeping Fido's Nose to the Grindstone

Of all a dog's senses, none surpasses his sense of smell. You might even say a dog sees his world through his nose, which provides him all kinds of information about the world around him, like who left the last pee mail on the neighbor's shrub. And we've all seen how dogs greet one another; it's not with a handshake. In addition to a dog's everyday life, in which he uses his nose for the standard sniffing and exploring, his valuable sense of smell also functions in count-less untraditional ways, including the work he does as a law enforcement professional.

A dog's nose is a million times more sensitive than our own. That's because he has forty times the number of brain cells for scent recognition humans do. The exceptional sensi-tivity of the dog's nose comes from having a larger olfactory area—130 square centimeters compared to our miniscule three square centimeters. Those cells are closely packed into folds or ridges that form a trap for capturing scent. They're actually visible if your dog ever lets you get close enough to peek into his nose.

Scent hounds like bloodhounds and basset hounds have long ears for a reason. As the dog sniffs along, those long ears sweep the ground and help to stir up the scent better for the dog to smell.

The dog's nose has a high capability for sorting out various scents, a skill that has not gone untapped by man. Because of their sensitive noses, dogs are even used for culinary purposes in the form of truffle hounds. The French poodle is often used to hunt for truffles. That gives a new meaning to the phrase, "Bone appetit!" We've also made good use of the dog's scenting ability by training them for security and safety purposes. Dogs have been trained to search for drugs and explosives, to detect gas leaks, and to track criminals over grass, dirt, and even cement or asphalt. Dogs can detect disturbances in the ground where a person fled from a crime and can even detect traces of skin cells shed on the trail that will lead the dog and the police right to the criminal every time.

In addition to their already-formidable scenting skills, dogs also have a vomeronasal organ in the roof of their mouths that enables them to actually taste a scent. What they are actually sensing are pheromones, which are associated with sexual attraction, but this sensitive organ is useful in other ways. Dogs can also sense fear or other emotions that are secreted from our bodies. What better skill to employ for the benefit of humans in sniffing out potential trouble?

Beyond Book Smart

Even dogs hate cell phones, as Snoopy, Kamal Shah's German shepherd, proved when she swallowed the Kenyan businessman's mobile phone. He looked high and low but couldn't find the phone until he heard a ringing sound coming from Snoopy. The dog underwent surgery to have the phone removed and both are in perfect working order. Can you hear me now?

Bomb-sniffing dogs have been a presence in airports for years, and they will most certainly be more visible now in train stations and other mass-transit areas in the wake of recent terrorist activity in London and places throughout the world. Fearing copycat attacks following the terror attacks in London on July 7, 2005, armed police and bomb-sniffing dogs boarded subways in Washington, D.C., New York City, and Chicago searching for any suspicious packages. These dogs provided invaluable assistance in securing areas where thousands of civilians go about their daily activities.

In these uncertain times, it's reassuring to know that our canine friends are helping to ensure our safety from security threats. Clearly, the benefits we derive from the dog's exceptional sense of smell are nothing to sniff at.

The Dogs of War

Men aren't the only creatures who make the ultimate sacrifice in American wars. Man's best friend has also offered his faithful service in K-9 combat units during wartime. In both World Wars and in Vietnam, scout dogs and combat trackers were on the front lines, risking their lives to save soldiers and civilians alike. Along with some of the brave men they served with, many combat dogs were killed in the line of duty. You may not see them featured as often on the evening news coverage of the engagement in the Middle East, but dogs are performing the same services in Iraq and Afghanistan as in previous wars. The bond that forms between a dog and his dogface handler is forged as strong as the steel of a bayonet.

Beyond Book Smart

Don Mobley of Anchorage, Alaska, bearly escaped disaster when he was charged by an enraged momma Grizzly. The bear was gaining on him when his three-year-old German shepherd mix, Shadow, intercepted the bear and kept it from attacking Mobley. For saving his owner, Shadow was rewarded with his favorite: pizza and potatoes. But no bear claw danish was served for dessert.

Many heroes answered the call to provide emergency services following the World Trade Center attack on September 11, 2001, and some of those heroes came running on four paws. The federal government sent out a nationwide call for the most highly trained canine search teams. Search Dog Foundation (SDF) responded, providing thirteen canine search teams to seek out survivors of the worst terrorist attack on American soil. Because of the rigorous training these dogs undergo with SDF, the U.S. Federal Emergency Management Agency–certified teams of dog and handler were able to perform under the most difficult and precarious conditions they had ever faced. Fortunately, no SDF dogs were injured in the search at Ground Zero because they and their handlers are trained to work on rubble piles and shifting surfaces and knew how to avoid dangerous spots. It's interesting to note that all the SDF dogs that did rescue work at the World Trade Center were themselves rescued from pounds and shelters and given a second chance at life doing important work. It's a canine career not without risk, but being a hero is never easy, for a dog or a man.

Canine teams are deployed in all kinds of disasters across the nation. In the aftermath of natural disasters such as Hurricane Katrina, dogs search for the lost and injured. SDF acquires dogs from shelters and rescue groups across the nation, looking for the dogs that have the qualities needed to make it through the rigorous training program and the demands of deployment. It's not a job for every dog, but it's a job dogs live to perform.

Honoring Canine Comrades

While dogs have been serving this country since World War I, a memorial was not erected for them until May 8, 2004, when a new pedestal was unveiled at Fort Benning, Georgia. In remembrance of the 400 dogs killed in action since the start of World War II, a boatswain's bell tolled at the reading of each name. Envisioned by SFC Jesse S. Mendez (USA retired), this pedestal is one of eleven others surrounding a larger War Dog Memorial, and more are planned in the future.

Jeffery Bennett developed the idea for a memorial and decided two memorials were needed. The fact that many of the military war dogs were abandoned in Vietnam motivated him to move the project forward.

The two War Dog Memorials are located at March Air Force Base outside Riverside, California, and at Fort Benning. The nineteen-foot-high bronze memorials, created by world-renowned sculptor A. Thomas Schomberg, depict a GI in combat gear with a vigilant dog at his side. The inscription reads: "They protected us on the field of battle. They watch over our eternal rest. We are grateful." Other pedestals in the ever-increasing memorial project identify K-9 combat units, and they have been funded and designed with the help of the respective units. These memorials may be a bit late, but it is nice to know that the canines that have died while serving their country are finally getting the recognition they deserve.

> "There is only one smartest dog in the world, and every boy has it."—Louis Sabin

The Healing Paw

Mary Myers says her Parson Russell terrier, Baby, is very attentive to her elderly mother, Dorothy, who has taken several serious falls. Every time Dorothy gets out of her wheelchair, Baby follows her. If Dorothy does something out of the ordinary, Baby will come to Mary and stare at her to indicate Mary needs to come and check on her mother. Mary says, "In every incident, Mom has been doing something that I usually attend to her in doing."

Mary's case is not unusual. Many dog owners report that their dogs not only look out for their general safety, they also dabble in the medical field. For example, my doctor doesn't wear scrubs, listen to my heart with a cold stethoscope, or dispense pills to cure my ailments. My doctor has miraculous healing powers, and I feel completely comfortable and relaxed in his presence. Best of all, I don't need medical insurance for his services. Why, he even makes house calls!

Like many dog owners, my physician is kind of furry and has four legs. My dog not only knows how to heel but also how to heal.

"In dog years, I'm dead."—Unknown

Studies have shown that just the act of stroking a pet has a beneficial effect on our health by lowering our heart rate, blood pressure, and respiratory rate. During a particularly stressful day recently when my blood pressure was in the stratosphere, I tested this theory on myself. I saw my systolic number on the pressure meter drop thirty points and my pulse rate plummet after petting my dog for just a few minutes.

When I was seriously ill with pneumonia in 1991, my dogs were instrumental in my recovery. Dolly, the angel basset I'll always believe came briefly into my life to help me through that difficult period, slept at my feet every night while I was sick. It's the warm memory of her and my other basset, Patience, basking contentedly with me in the sunny front room of the house during my convalescence that comforts me now that they are both gone.

When my father suffered a debilitating stroke several years ago, it was the presence of Laddie and Duffy, the two little Scottish terriers my parents had recently adopted from the pound, that was instrumental in his rehabilitation. Dad had lost all feeling in his right hand, the one he used to strum his beloved guitar. Laddie, the dog that bonded most strongly with my father at the outset, sensed that Dad's right paw was hurt. He licked and licked the injured hand, and before long the feeling returned fully enough for Dad to play his guitar again. I'm certain that it's nothing less than a tail-wagging miracle that my father is still with us today.

> "I've caught more ills from people sneezing over me and giving me virus infections than from kissing dogs."—Barbara Woodhouse

Many other people credit their survival of life-threatening diseases to the healing presence of a pet in their lives. This phenomenon is why more nursing homes, convalescent hospitals, and rehabilitation facilities are adopting pet therapy programs. Pets reduce stress and encourage social interaction, and their presence alone makes people happier. The contact we have with our pets makes them happier and healthier, too. The simple act of grooming a dog is soothing to both you and your dog. Petting, grooming, and loving your dog is nothing less than a win-wag situation.

Our dogs give us unconditional love. They make us laugh. What is more healing to body and soul than love and laughter? For some people, dogs have had a far more beneficial effect on their overall health than pills have, and there are never any side effects. Sure, we still need medicine for certain illnesses, but the exercise you get from taking your dog on daily walks can have amazing curative effects on many health issues such as obesity, heart disease, and diabetes.

There's nothing that can raise me up like Lazarus as effectively as when Dr. Bubba comes to my bedside, nudges my hand with his cold, wet nose, and gives me the look that says, "Okay, Mom. Hurry up and get well so you can walk me again." No bedside manner I've encountered in any medical professional can compare to his.

Dr. Bubba may not have a license to practice medicine, but he always listens to my heart and helps cure whatever is bugging me. I'm still waiting for the day when I hear a medical doctor say, "Pet a dog and call me in the morning."

"My little dog. A heartbeat, at my feet."—Edith Wharton

Doggie Diagnosticians

Biodiagnostics is a relatively new and unknown branch of medicine that uses animals such as sniffer dogs to diagnose diseases in humans. Since 1989, there have been three cases in the United Kingdom in which dogs detected skin cancer in humans. The dogs behaved strangely, persistently sniffing a lesion on their owners' bodies, which caused their owners to seek medical attention. In every case, the lesions were found to be in the early stages of cancer. Two of the cases were malignant melanomas, conditions that if diagnosed too late can spread throughout the body and prove fatal.

Since the first successful doggie diagnosis in April 1989, dogs have been trained to recognize cancer with astonishing success, but more work needs to be done to harness this amazing skill. If more studies are forthcoming, dogs may eventually be trained to sniff out a broad range of diseases in humans. It's been determined that certain diseases emit a distinctive aroma that a dog's ultrasensitive nose can easily sense. For instance, if you have diabetes, you may emit a sweet smell on your breath. If your dog sniffs persistently at a mole or some other lesion on your body, it might be wise to pay attention to that behavior and seek medical advice. It could save your life. Then again, if Spot persists in smelling your breath, he might just be trying to figure out whether that pizza you ate was sausage or pepperoni.

> "Dogs are wise. They crawl away into a quiet corner and lick their wounds and do not rejoin the world until they are whole once more."—Agatha Christie

Recently, in Wales, a Parson Russell terrier named Milo saved his master's leg from being amputated by licking the affected limb for hours at a time. The dog's owner, Mitch Bonham, was in danger of losing the leg after a heavy anchor fell on his foot when he was a diver in the Royal Navy. His leg atrophied, and his doctor told him that the entire limb would have to be removed if it didn't improve. That's when a canine consultant was called in. As Bonham lay in bed, Milo showed keen interest in the injured foot and kept trying to lick it. At first, Bonham tried to keep the terrier away, but finally decided that it couldn't hurt. He was probably going to lose the leg anyway.

One day, he felt his toe twitch, and the feeling began returning to his leg. Apparently, the constant attention Milo had given the leg had stimulated the nerves and helped carry oxygen to the affected area to help it heal. Before long, the tenacious terrier had the problem licked. Bonham's leg was saved by his faithful Milo, whose payment for services rendered was a juicy bone that was bigger than he was.

As we learn more about the scentsational powers of the dog's nose and the exceptional abilities some dogs possess to detect disease in humans, early detection of life-threatening illnesses may be just a sniff away.

Dogs At Your Service

We've all seen these working dogs in public, assisting their owners with daily tasks many of us take for granted. You'll see them guiding the blind or walking beside a wheelchair. They are called Canine Companions for Independence (CCI), and that's what they provide: independence for the physically challenged.

An intensive training program begins when the pups are two months old, at which time they are taken from the homes of volunteer caretakers. The pups are brought to a special CCI campus located in Santa Rosa, California. After a health clearance, they are placed with volunteer puppy raisers. You may have noticed dogs wearing special "Dogs at Work" jackets, an indication that you shouldn't pet them or distract them while they're working.

When the dogs are about a year old, they return to the center to complete their training. Once they complete their advanced training program, they are ready to meet their human partners, who must also undergo a two-week intensive training period. The end of team training is celebrated with a graduation ceremony to mark the end of a job well done, but the real job is just beginning in a long-term relationship between the CCI dog and his owner. The relationship with CCI doesn't end there, though. Two-legged and four-legged graduates keep in regular touch with CCI through special follow-up programs.

Even if a dog flunks out of the CCI program, sometimes there's still a future in helping people in nontraditional ways.

Stayin' Alive

Agility is about surmounting a course of obstacles. Virginia Gilmore, of Princeton, Texas, faced an obstacle that threatened to take her permanently off course when she was diagnosed with breast cancer at forty-seven. Fifteen years later she's going strong, thanks largely to agility and Sooty, an Australian cattle dog cross with an equally feisty attitude.

"The year of 1989 was not a good year," Gilmore says. Overcoming the hurdles of her husband's skin cancer surgery, her father's death, and a mastectomy took a leap of faith. Agility became her salvation. "I saw my first agility trial in 1990 at a local AKC breed and obedience trial," she says. "I came, I saw, I was hooked!"

She first tried agility with Treat, an Australian shepherd-Border collie cross that was no treat to train. "She taught me more about dog training than all the others put together!" Gilmore says. "She may have had lots of issues, but she sure could give good hugs." Then came Chili, an Aussie with stage fright. "If she wasn't having fun, then I wasn't having fun." Chili now rests on her laurels, and the couch, with her current title, Master of the "Nub Rub Boogie."

Third time was a charm when Gilmore rescued five-week-old Sooty from a dumpster. One person's trash is another's treasure. "As soon as she was old enough, I started training her, first in obedience classes and then in agility," she says. Who rescued whom?

"The very worst thing about having cancer is that it takes away your daydreams. Suddenly, your future appears very bleak and uncertain. I was afraid to look forward to next year." Gilmore says. "But after getting involved in agility and dog sports, I could at least look forward to practice next weekend. Later, I began to be able to look forward to a trial next month, and then eventually to next year at Nationals."

For a decade, she and Sooty competed coast to coast, making friends and memories along the way. "When I'm sitting in my rocking chair in 'the home,' I'm sure I'll be boring all my fellow residents with 'I remember the time . . .'" Like the time she fell on her butt in front of the standing-room-only crowd at the USDAA Nationals in San Antonio. "I still argue with Tim Laubauch that Sooty should not have earned that refusal." She fondly remembers making the finals in the 1996 USDAA Grand Prix in Ventura, California. "So what if she went off course on the fourth obstacle—we made it to the final round! Whee!"

Anyone who's ever had an illness make them sit, stay, and heal appreciates a dog's nursing nuzzle. "They don't care if your body is scarred or disfigured or if you're having a bad hair day. They're always there for you," Gilmore says. After her husband died, the dogs were the one constant in her life. She jokes about Chili, her primary consoler: "Nothing like a big, fuzzy-butted Aussie to hang onto."

Sooty retired at age eleven with a string of titles as long as an agility course tunnel. These days, Gilmore herds with Wish, a two-year-old Border collie. She is also training Shine, a four-year-old Aussie-Border collie cross rescued from a local shelter, for competition in obedience and agility. "I should have had Shine when I was about twenty years younger and thirty pounds lighter," she says. "She's a handful—fast and very smart, and we're coming along, but it's been a real challenge for a lady 'of a certain age.'" Small challenge compared to surviving cancer.

"I was one of the lucky ones; my prognosis was very good from the start. Many cancer specialists seem to believe that a positive attitude and an 'I'll beat this' spirit is one of the vital forces in long-term survival of the disease," Gilmore says. "The competitive spirit that gives people the drive to succeed in any sport will serve them well in what may be, literally, the fight of her life."

Early diagnosis is crucial. "While we tend to rush the dog to the vet for the slightest hint of problems, we tend to ignore our own health," she says. "A mammogram probably saved my life." Women too often avoid them, but until breast cancer screening becomes a less pressing matter, it's mammography or bust. "Yeah, it's uncomfortable," she admits, "but a mastectomy is no piece of cake, either!"

"Use sunscreen," she also advises. "Since agility is such an outdoor sport, we all need to be very aware of the dangers of overexposure to the sun and use appropriate precautions." Her fair-skinned husband never used sunscreen and rarely wore a hat. "Folks, let me tell you, you won't be running agility for quite a while after having a skin graft!"

"Life is too important to be taken seriously," Gilmore quotes Oscar Wilde. "Sometimes I worry that we are taking agility too seriously, making it too technical, always trying to cut that nanosecond of time, making it difficult to feel good about myself unless I've had that perfect run and blasted the competition," she says. "It's more important than that—it's about good times, good friends, and good dogs!"

> "I've been on so many blind dates, I should get a free seeing-eye dog."—Unknown

Above and Beyond the Call of Duty: Extraordinary Dogs

While any dog's accomplishments can be extremely impressive to his owner, some dogs truly perform extraordinary tasks—especially when their owners are facing a difficult time. Physical handicaps may hinder a human's ability to function in society were it not for aid dogs. Sadness stemming from the death of a loved one or other tragedy may be too much to handle without canine companionship. This has never been more clearly evident than with the reluctance of evacuees to be parted from their canine companions amid the devastation of Hurricane Katrina. Many refused to leave their dogs behind to save themselves. When you've lost so much, your dog may be all you have left. Perhaps in the future, better arrangements can be made so that people who are already traumatized and displaced need not abandon their pets or fear for their survival in such disasters.

In the following cases, dog owners faced feelings of helplessness and desperation and may have fallen into a depression had it not been for friendship in the form of a dog. Regardless of how these dogs rate in terms of IQ, it is apparent that there are more important canine traits than traditional smarts.

Overcoming Disability in Agility

When Mike Penketh approaches the starting line at an agility competition, people often stare. You might think that's because Mike has no hands, but it's really Magy, the pretty blonde girl at his side, who is the object of all the attention. Magy is Penketh's golden retriever service dog.

Magy was a release dog from Canine Companions for Independence (CCI), which means that although she went through all the training, she didn't graduate from the program. Maggie's habit of chewing her toes may not be as bad a habit as Walter's, the hero of the children's book *Walter the Farting Dog,* but her foot fetish kept her from becoming a card-carrying CCI dog. "If they gave that dog to a quadriplegic in a chair, he couldn't take care of her," Penketh explains. Magy serves as his hands. "She'll carry things for me, carry my car keys for me, get the paper for me, get my slippers for me. She's my twenty-four-hour-a-day companion."

Ordinarily, Penketh would have faced a two-year wait for a release dog. "I asked about a flunkie dog and was firmly corrected—CCI doesn't have any 'flunkie' dogs," Penketh says. But they do have "change of career" dogs. Fate kindly intervened, and two months later he had Magy. Man and dog were about to embark on a new career.

A former Marine Corps pilot, Penketh lived life at top speed. He flew a jet in Vietnam and admits he's doggoned lucky to be around to tell about it. When his jet crashed, he sustained a back injury, but that didn't slow him down much. Years later, following rehab from back surgery, he won the 1993 Reno Air Races and a biplane championship four days before disaster struck again.

"I was at the Bonneville Salt Flats in my race car," Penketh says. "The car crashed at almost 300 miles an hour." He lost both his hands in the accident. That meant he wouldn't be racing any more biplanes or high-performance cars, but it didn't mean he was out of the races for good. He soon became involved in competition of another kind.

"I was watching Animal Planet on TV, and I saw dog agility on there. I thought, 'That looks like fun!'" In the beginning, he tried to get his petite wife, Mary Ann, interested in this burgeoning dog sport, but it wasn't exactly her cup of kibble. "Mary Ann didn't seem interested, so I took a couple of agility lessons myself to find out about this sport. We trained for about two years before we entered our first trial."

Penketh and his golden girl took lessons from Marna Obermiller of Precision Chaos in Dixon, California, near his lifelong home of Vacaville. "She discouraged us from competing until she thought we were ready. In hindsight, it was a very wise move. You don't want to go out and beat your head against the wall." He says at first he used to get lots of looks because of his disability. "No one even looks at me any more 'cause they all know me."

These days, Penketh never misses a Dixon competition or other trials around northern California. Magy took to agility better than dogpaddling in the Penketh's swimming pool. Magy loves to go "jump-jump." Penketh says, "The dogs don't find anything difficult. You have to learn to handle the dog, send proper messages to the dog. He looks at your hips, shoulders, eyes, hands, interprets all those movements. People develop a tremendous teamwork with the dog."

Teamwork has paid off for Magy and Penketh. "She's gotten about twenty titles in our second full year of competition," he says. "In AKC she competes on the Excellent level; in NADAC she competes on the Elite level. The third organization we compete in is Canine Performance Events, where Magy is in Level Five." One might attribute their success to the fact that Magy and her handler are so perfectly matched. A service dog is specially trained to be watchful of his master's every move and nuance of expression. This is ideal for Penketh, who can't use hand signals to command Magy.

Besides agility, Penketh and Magy share another career in a program called Touch of Understanding. Twice weekly, they talk to schoolchildren, doing over 100 presentations a year. Magy is the star of the show. "She goes with me, wears a little cape, and carries my video tape. I talk to kids about disabilities, and I always throw in a ten-minute session talking about service animals and proper etiquette when you see a service animal. Magy really adds a lot to that."

Magy isn't Penketh's first golden retriever, but she's his first service dog. "She opened up an entirely new chapter in my life," Penketh says. "She gives me a lot of pleasure, makes me smile all the time." When asked if competing in agility presents any particular challenge for him, Penketh laughs. "The biggest challenge I have is eating lunch."

If you should happen to see Mike lunching with a gorgeous, curly-haired blonde with a winning smile, her name is Magy.

Bowie Bows to None

"Bowie rocks!"

Irene Culver, one of the volunteers at the Placer SPCA, says they aren't talking about the singer with one blue and one green eye. They're talking about Bowie, a pit bull whose eyes are also mismatched, who is leaping over the high jump in the shelter's new agility course. It's a training session for dogs and volunteers to prepare for the Roseville California's weekend Adoptathon.

No one had expected Bowie to participate in the training, much less compete. A few weeks before, visitors to the shelter might have labeled the snow-white, well-muscled dog as autistic; he simply appeared to lack any social skills. But that wasn't the problem—Bowie is deaf.

A volunteer brought in a vibrating collar and began working with him. "First," says Matt Green, Placer SPCA operations manager and animal behavior specialist, "she had to get his attention. The collar worked." Then Bowie learned and obeyed hand signs. Success changed both Bowie and the way people looked at him, but the biggest breakthrough came when the volunteer let him try the agility course. Bowie liked the obstacles and proved himself a natural talent right away.

Entry-level agility trainers learn how to train formerly unadoptable dogs at this animal shelter, which attracts hundreds of volunteers annually. One important goal is to teach and involve the surrounding community, and another is to reach higher adoption rates. Agility training is one of the primary ways they meet both goals.

Dogs that spend most of their time in shelter kennels are not likely to be at their best when potential adopters come to see them. Green notes, however, that the dogs are more lively and eager to interact with visitors when involved in agility training. "It helps increase their confidence," he says, "and a confident dog is more attractive and more likely to be adopted."

At a workshop for volunteer agility trainers, Green passes around a list so they can write down their choice of dogs to handle at the competition. Then the group moves to the agility course, with Green's own dog, Snap, and with Bowie, to demonstrate how to train the dogs. The volunteers learn the ins and outs of the course as they watch Snap breeze through the course, even making it over the most difficult part of the course—the teeter-totter.

After Snap makes it through the course, volunteer Edith Nicoletti takes Bowie through for his turn. The dog seems to be laughing and having a great time, until he spies his own shadow. Then, as if the world has disappeared, Bowie focuses on the dark shape, nose to the ground, still as a statue. Volunteers and visitors to the shelter who have gathered to watch are amused by the dog's total concentration and complete disregard for Nicoletti's hand signals as she tries to get him back on course, explaining to the audience that Bowie is deaf. When she does get his attention, she demonstrates the hand signals he understands. She shows the dog two closed fists, and then points over his head. Bowie sits. The visitors applaud, and though the dog can't hear, he seems to know his obedience is appreciated.

Edith Nicoletti leads Bowie back to the shelter. She says she feels he is ready to compete and will do well. But as fate had it, he does not compete during the Adoptathon—a few days before the competition, Bowie is adopted.

"Bowie," said Matt Green, "had virtually no chance of being adopted before he was trained, because he's deaf and because of people's misconceptions about pit bulls. But the agility training uncovered his sweet and fun-loving nature and made him attractive."

Our intelligent, faithful dogs help us in so many ways that it would be impossible to cover everything they do for human beings within the scope of these few pages. No book about dogs would deny the unique bond we have with these animals and the myriad positive ways in which they impact our lives. Our symbiotic association with the canine species is the best doggoned relationship man has ever had. They guide us, they rescue us, and they comfort us. There are countless miraculous stories about how dogs have helped their owners to thrive and survive. Every day, our own dogs may surprise us, whether on a large scale, like the dogs featured in the stories in this chapter, or with something as small as an extra friendly lick.

Chapter 10

On Becoming Fully Canine

"In order to really enjoy a dog, one doesn't merely try to train him to be semi-human. The point of it is to open one-self to the possibility of becoming partly a dog."—Edward Hoagland

The newest generation of Yuppie puppies are reaping the rewards of being the apples of their owners' eyes. Dogs have graduated from the role of home sentry to commanding the center of attention in our lives, and we spare no expense to ensure that our dogs are happy and well cared for. Like helicopters, pup parents hover over their charges and are constantly searching for new and better ways to keep their canine companions engaged and entertained. Some people even change careers just so they can spend more quality time with their dogs, but I'm not mentioning any names.

What Dogs Are Doing

Dogs of every breed and age are enjoying a broad range of activities that are benefiting not only the dogs but also their owners. Contrary to the old adage, you *can* teach an old dog new tricks. A dog's trainability has more to do with his breed than his age. In fact, keeping your old Bowser engaged in learning a new task that is not beyond his physical capabilities can improve the quality of his life and even prolong it. The following are just a few of the many fun pursuits people are discovering with a little help from their four-legged friends.

Agility in Action

One of the more popular new doggie sports, and one that boasts legions of devotees, is the sport of agility. My introduction to dog agility actually occurred decades ago in a grade school auditorium when Nehi, the Wonder Dog, and his trainer came to Woodlake Elementary School to teach us children about safety. Nothing could have been more exciting to the children than to see a dog at school. We knew it was against the rules to bring a lamb, but we got a dog! Nehi coming to visit us at assembly that autumn day was the next best thing. To see this dog accomplish with the greatest of ease feats of amazing dexterity and obedience was a thrill beyond compare. The clever tricks the terrier performed weren't called agility in those days, but that's what he was doing.

Beyond Book Smart

A small white Chihuahua named Spicy became a Robinson Cur-rusoe when she was stranded on an island for seven days after falling or jumping from a pontoon boat. The little castaway survived her ordeal by eating insects and small rodents. That was sure more than a three-hour tour, little buddy!

Nehi, named not only for his short stature but also for a popular soft drink of the day, was a wire fox terrier and a dervish of energy, as is typical of the breed. Leaping, whirling, flipping—he seemed to fly across the auditorium, as though a pair of wings had sprouted through the dense coconut mat of his wiry coat. Thanks to that lively terrier tutor, we never forgot the safety lessons he taught us that day.

Some dogs are not by their nature or design well suited for certain activities. For instance, you aren't very likely to see a basset hound or other dogs with short legs race on an agility course, but sometimes you just can't keep a good dog down (even though he's already pretty low to the ground).

The Lowdown on Short-Legged Breeds and Agility

Eager yips and barks disrupt a tranquil Saturday morning in the northern California town of Dixon. The AKC agility trial at the fairgrounds is underway, and it's Penelope's turn to run the open jumpers and weaves course. She is the only basset hound competing at the trial today, and many might wonder why she's here. After all, bassets are just stubborn couch spuds that can't be trained to do anything on command besides eat. They don't actually run, do they? Little does anyone suspect how agile and speedy these belly-draggers can be when they've a mind to. Molly's Sweet Penelope, who at five years of age boasts a string of titles longer than her ears, is about to prove that not only can bassets run but they can run doggoned fast!

Of course, these stubborn, waddling, slo-moes have also elicited a few snickers over the years, but Penelope is not your average sofa-warming basset hound. "Penelope usually has the fastest time in her division," says Shirley Harris, her owner and trainer. "She really is a fast hound!"

Harris disputes the perception that bassets are any more difficult to train than other breeds. "I actually enjoy training more than trialing," she says. "It's really satisfying to see my dog learn new skills." Harris became involved in agility because it was a way for her to spend quality time with her dog. "I had seen agility trials on television and thought it looked like a lot of fun, not just for the handler but for the dog," Harris says. "Because I was new to the sport, I really had no idea what made for a good agility dog. As it turned out, Penelope does have a lot of drive."

Harris found that some kinds of agility obstacles were harder than others for Penelope to master. "The teeter-totter and weave poles were the most difficult," Harris explains. "Penelope didn't like the noise and motion of the seesaw." The weave poles are especially tricky for long-backed dogs because of the slalom action required to execute the quick turns. These dogs may remind you of kids' Slinky toys of the 1950s, but they don't flex the same way.

Beyond Book Smart

Pooh Bear, a thirteen-year-old, six pound Pomeranian, traveled for six years, covering over 1,200 miles to find owner Bambi Lesne after being abducted from her home in Panama, Florida. Dirty and bedraggled, she was found in Cincinnati, Ohio, and reunited with her adoring owner.

As for those voluminous ears, they present no impediment to competition. The snood, which keeps basset ears tidy at Westminster, is never worn at agility trials for safety reasons.

Short dogs can jump, but they can't jump high. You aren't likely to see a basset competing in the flying disc. "I show her in the lowest possible height allowed," Harris says of Penelope.

As with all areas of agility and dog training overall, moderation is key to preventing injuries. "I trained her on all of the agility equipment; however, we do not train often and when we do train, we do not train for long periods of time," Harris says. Despite precautions, injuries sometimes occur. Penelope sustained a back injury but had physical therapy and is doing fine. Not so for Yum-Yum the Dachshund, whose agility action days are numbered, according to Linda Goodwin, who says an X-ray of the dog's spine showed there might be a problem.

"Agility has allowed me to become better educated about dog training," Harris says. "It is important to focus on your relationship with your dog and become the primary reinforcement." Harris and Penelope's seamless teamwork attests that their relationship is as solid as a Dentabone.

Doggie Artists: What Dogs Are Doing to Keep Busy

Nowadays, dog parents worry as much parents of human children about keeping their fur kids happy and entertained. Are we raising a generation of spoiled canines with entitlement issues? One can only hope. Dogs are due a lot of payback from us humans for all they have given us over time. In the past, a dog was something you kept tethered at the end of a chain and tossed a bone every now and then. Today, however, it is evident that more people look upon dogs as cherished members of the family. In the bargain, we have discovered that spending more quality time with our dogs is as beneficial for us as it is for the dogs. There is a broad range of activities for dogs that even Timmy never dreamed of doing with Lassie, although her silky paws would have made great paintbrushes for dog art, just one of the many new hobbies dogs are getting to enjoy these days. And they don't even have to retire first. The following are just a few of the fun pursuits for people and their four-legged friends. These examples may leave you wondering what's next in doggie amusement, Canine Carnival Cruises or Six Wags.

Dances with Dogs

Unlike the recent hit television show, *Dancing with the Stars,* you may not see this competition on prime time television, but everywhere across America people are tuning in on the fun of the latest canine musical/obedience freestyle craze, otherwise known as dog dancing.

Your partner may not wear a tux and top hat, but he can tango and foxtrot every bit as well as Fred Astaire or Patrick Swayze, even though he has two left feet. If your dog hasn't had a bath recently, you might even do a little dirty dog dancing.

Just as in ballroom dancing, the dancing teams are scored on technical and artistic elements of their routine. They are also judged on their costumes and the style and content of the dance. Well, maybe these dancing partners are sporting a tux and top hat after all. They're already wearing tails.

This sport has grown by leaps and twirls since the early 1990s when it first caught on. It emerged out of agility and other canine competitions when owners began adding dance elements to add a little pizzazz to their routines. In 1999, Patie Ventre, a former ballroom dancer who was very interested in dog sporting events, founded the World Canine Freestyle Organization. The rest is dog dance history. Cha, cha, cha.

Beyond Book Smart

Sam, a Chinese crested whose hairless body and crooked teeth helped him win the title of World's Ugliest Dog at the Sonoma-Marin fair three years in a row, died just days short of his fifteenth birthday. His owner, Susie Lockheed, loved him despite his ghastly appearance. Guess Sam must have had some big brains, or, "a great personality."

Vincent van Dogh

Dogs have been inspiring great artists for centuries, but these days there are dog artists of a different kind. These painters don't create their art with paintbrushes but with paws. Miraculously, some of the artworks are commanding higher prices than many of the old masters ever received for their work in a lifetime.

Dog art is offered at many doggie day camps and day-care centers, but most avant-garde artists of the canine persuasion prefer to work out of their home studios. The best artists are already equipped with the perfect brushes for grabbing the most color off the palette at once—a tail. Just dip the fine dog-hair brush in the paint, apply the swishes and flips of paint across the canvas, and voilà! You have your own Jackson Pawllock. In fact, some brand new art forms are emerging in the dog art world, such as Pointer-illism and Pup Art.

Splish, Splash

Some dogs take to water more readily than others. In fact, it's a challenge to keep some retrievers out of the water. If you happen to have a swimming pool, your dog is likely to spend more time frolicking in it than your kids. Fortunately for your water-loving canine, there are some new activities you might like to try that will have him dogpaddling for joy.

Dock dog training is one of the newer dog sports afloat these days, and it's becoming a big splash with dog owners and their pets. In this sporting event, featured on ESPN's *Great Outdoor Games,* dogs race to the end of a dock, then fly off the end into the water. A toy or other object is thrown into the water to encourage the dog to make the leap. Distance and style are the aim, as well as how well the owner works in concert with the dog. Dock Dog events are held all over the country and are open to dogs of any size or breed, including mixed breeds. While these events are intended to be fun for dogs of all abilities, participation in the ESPN events are by invitation only.

Hang Ten with Your Best Friend

If your dog craves adventure on the high seas, you can always take him on a surfing safari and ride the waves with your Moondoggie. In *The Dog's Guide to Surfing*, author Danny Galligini tells you everything you need to know about teaching your dog to be a Big Kahuna canine. The book includes stories about surf dogs as well as lessons and advice and features bitchin' gear and surf wear for your board buddy. Galligini claims your dog may get so hooked on hanging four paws on the longboard that he may want to join you on your snowboard, wakeboard, or kiteboard. Whatever the board, your dog won't be bored.

Dogma—What Can Our Dogs Teach Us?

We spend a lot time teaching our dogs to be better dogs and to fit into our concept of what we think the world should be. Sometimes, we forget that dogs also have many lessons to teach us about being better people.

Dogs manage to start their day without the use of caffeine or pep pills. They are always cheerful, no matter how they might be feeling, and they don't complain about their aches and pains or bore others with their troubles. They eat the same food day after day and are grateful to receive it. They always understand if you're too busy to give them your attention and overlook it when people take things out on them when, through no fault of their own, things go wrong. They can take criticism and blame without holding a grudge. Dogs are never deceitful. They can relax without liquor and sleep without the aid of drugs. There are few humans who could say the same. Clearly, we have much to learn from our dogs.

When is the last time that you got in touch with your inner dog? I'm speaking of tapping into those qualities of forgiveness, love, acceptance, simplicity, and joy that we possess but that seem to be in ever-shorter supply on the planet. All these traits are innate in our dogs. Perhaps we're the ones who need the most training.

We know these "doG"-given abilities lie buried inside us somewhere, but like a dog that searches for a prized treat, we might need to dig a little deeper now and then to uncover them. To my knowledge, there is no Daisy Hill People Farm for a brush-up course on becoming fully human. Still, sometimes the lessons we most need come to us when we least expect them.

Recently, while sitting in the garden with my dog, I happened to witness the annual migration of painted lady butterflies. I realized while observing this miracle how separate we have become from the natural world in our daily lives. We exist in ever-shrinking spaces confined by concrete and steel, besieged by sights and sounds that are alien to our primitive senses. The more mechanized life becomes in the twenty-first century, the more we long to reconnect with nature. If we can't escape the constrictive cocoon we've spun for ourselves, at least we can regain some of Eden's serenity and innocence in the company of our beloved pets, as so many dog owners do.

Pet ownership is at an all-time high, and we'll spend billions of dollars on our pets every year to provide them with the best food, shelter, and sundry comforts money can buy. When asked what our pets provide us in return, we respond with words like love, loyalty, and companionship. Research has shown they also provide us with significant health benefits, but they give us even more. Our pets provide a way for us to reconnect with the world and with one another, whether they are a Lassie or a Scooby Doo.

For example, do you know your next-door neighbor's name? His pet's name? If you're anything like I am, you learn the name of your neighbor's dog before that of your neighbor. You and Rover have no doubt had long conversations through the back fence. Let's face it: Rover is probably more approachable and less territorial than his owner is. One day you happen to cross paths with your neighbor on the street while walking your dogs. Despite your best efforts to avoid each other, the dogs entwine their leashes around you both in a maypole dance of unbridled enthusiasm. Standing there nose to nose like your dogs, it's hard for the two of you to remain strangers. You discover that Rover's dad actually has a name and that you both have quite a lot in common, starting with your shared affection for dogs. You've connected. Amazing!

Things I Have Learned From My Dogs

- Sometimes just sitting quietly by someone's side is the best kind of comfort you can offer.
- Going for a walk is one of the greatest pleasures of the day.
- There's no reason not to wake up happy, and walk around with a bounce in your step.
- Every meal is pure heaven.
- People react much better to a friendly, happy, I'm-so-glad-to-see-you greeting than they do to, "What time will dinner be ready?"
- Patience, patience, patience.
- Lying in a nice spot of sunlight is a really fine way to spend an afternoon.
- People don't like to kiss you if your nose has been somewhere it shouldn't be.
- Loyalty is not a sign of weakness.
- Sometimes our smallest friends are our staunchest protectors.
- The downward dog stretch feels really good after a nap.
- Spending time with your loved ones is pure bliss.

The Tail End

Whether your dog tested at the level of a canine Einstein or a dog-school dropout, I'm sure we'd all have to agree that dogs all rate at the top of the scale in their ability to give us unconditional love. And what is more important than that? If they were to test us on this ability alone, we humans would fail miserably.

At the dawn of our association with the canine species, we invited these scavengers into our circle for the basic services they could provide to us: protection, companionship, and a little light cave cleaning. In return, we allowed them to sleep by the fire and tossed them a juicy wooly mammoth bone every now and then. Over time, that relationship has evolved to the point that it's often hard to tell who is more devoted to whom, man or dog.

In the final analysis, perhaps the smartest dog is the one that knows how to make us do everything for *him*. If this were the only measure of a smart dog, *all* dogs are geniuses. If aliens from another galaxy ever land on this planet to study the behavior of Earthlings and how we interact with our dogs, it would be immediately apparent to them which of the two life forms is the more intelligent.

"All of the good things that have come to me have come through my dog."—A dog owner overheard in New York's Central Park

Other Sources of Information

Organizations

Agility Action
www.agilityaction.com

American Kennel Club, AKC Headquarters
260 Madison Ave.
New York, NY 10016
212-696-8200
www.akc.org

American Pet Products Manufacturers Association
255 Glenville Rd.
Greenwich, CT 06831
203-532-0000

Camp Bow-Wow
www.campbowwowusa.com

Canine Companions for Independence
P.O. Box 446
Santa Rosa, CA 94502
800-572-BARK
www.caninecompanions.org

Greyhound Pets of America
800-366-1472
www.greyhoundpets.org

Guardian Angel Basset Rescue, Inc.
P.O. Box 288
Dwight, IL 60420
www.bassetrescue.org

Hemopet, Blood Blank Office
11330 Markon Dr.
Garden Grove, CA 92841
714-891-2022
www.itsfortheanimals.com

The Humane Society of the United States
2100 L Street, NW
Washington, DC, 20037
1-800-HUMANE
www.hsus.org

PETCO
www.petco.com

PetSmart
www.petsmart.com

Search Dog Foundation
206 N. Signal St., Suite R
Ojai, CA 93023
888-459-4376
www.searchdogfoundation.org

United Animal Nations
P.O. Box 188890
Sacramento, CA 95818
www.uan.org

United States Dog Agility Association
P.O. Box 850955
Richardson, TX 75085
972-487-2200
www.usdaa.com

War Dog Memorials
www.war-dogs.com

World Canine Freestyle Organization
P.O. Box 350122
Brooklyn, NY 11235
718-332-8336

Print Resources

Animal Fair Magazine
545 8th Ave.
Ste. 401
New York, NY 10018
www.animalfair.com

The Bark (magazine)
2810 Eighth St.
Berkeley, CA 94710
877-227-5639
www.thebark.com

Dog Fancy (magazine)
3 Burroughs
Irvine, CA 92618
949-855-8822
www.dogchannel.com/dog/magazines/dogfancy

How Dogs Think, by Stanley Coren. (New York: Free Press, 2004)

The Intelligence of Dogs: Understanding the Canine Mind, by Stanley Coren. (New York: Free Press, 1994)

About the Author

SUE OWENS WRIGHT is a writer of both fiction and nonfiction about dogs. She is a fancier and rescuer of basset hounds, which are frequently featured in her books and essays. Sue is a five-time nominee for the Mighty Maxwell, awarded annually by the Dog Writers' Association of America (DWAA) for the best writing on the subject of dogs. Her first nomination was for *Howling Bloody Murder*, the debut novel in the Beanie and Cruiser mystery series. She won the Maxwell Award in 2003 and received special recognition from the DWAA in 2004 for the Humane Society of the United States Compassionate Care Award. She won her second Maxwell Award in 2005. For more information about the author, please visit *www.beanieandcruiser.com*.